TURNING POINTS IN
WESTERN TECHNOLOGY

A Study of Technology, Science and History

D1193334

D. S. L. CARDWELL

Turning Points in Western Technology

A STUDY OF TECHNOLOGY, SCIENCE AND HISTORY

Science History Publications

A DIVISION OF
Neale Watson Academic Publications, Inc.
NEW YORK

Science History Publications

A DIVISION OF

Neale Watson Academic Publications, Inc.

156 Fifth Avenue, New York 10010

ISBN Cased edition 0 88202 004 8

Paperback edition 0 88202 003 X

Library of Congress Catalog Card Number 72–75374

3rd printing 1976

Printed in U.S.A.

Contents

Introduction

It is, in view of the immense importance of technology in the modern world, astonishing that more attention is not paid to the study of its history. In part this comparative neglect is due to the apparent complexities of the subject. Virtually every familiar object of everyday life includes or implies a whole series of elaborate inventions. A television set is an obvious example; but even the simplest utensil or tool which has changed very little in outward form over the years is now often made of entirely new substances and manufactured by means of highly sophisticated machinery. In short it would almost seem that a history of technology must be no less than a history of all aspects of material culture. But further thought suggests that this complexity is not so much the cause as a consequence of the undeveloped state of the subject. Any history necessitates a series of discriminatory judgments: what is held to be unimportant is ignored or relegated to the enthusiasts and antiquarians.

What then are or should be the common talking points of the history of technology? Before we suggest how this question might be answered let us glance at a few of the answers which have already, in effect, been given. The first thing that must strike us is the surprising extent to which the history of technology is still influenced by the writings of Samuel Smiles, whose famous *Lives of the Engineers* was published about one hundred years ago. Smiles wrote about those commanding figures—Brindley, Smeaton, Watt, Boulton, Rennie and Stephenson—who by moral and physical courage, strength of character, vision and aptitude transformed Britain from what was, to a great extent, an undeveloped wilderness into a modern industrial country. Smiles, in fact, was not concerned with other countries; he did not attempt to relate technology to other activities, such as science, and he did not probe into the nature of technological invention, or the methods of technology. It is a matter for regret that Smiles apparently made no contact of any sort with his equally distinguished contemporary, the first

vii

English historian of science, William Whewell.* It is indeed doubtful if Smiles could have understood Whewell and his ideas but the converse would almost certainly have been untrue.

There is today probably more interest in the history of technology and in the history of science than at any time before, including that period when Smiles was writing for his enthusiastic public: the early Victorians who had been so impressed by the visible effects of industrial revolution which they saw all around them and which they had celebrated in the famous exhibition of 1851. Modern interest in the history of technology shows itself in at least four different ways.

In the first place there is the enthusiasm for industrial archaeology. The stimulation for this comes from the growing recognition that the achievements of early industrial Britain were truly unique and that many valuable relics were either falling into decay or in danger of being swept away: these should be preserved for posterity. This is excellent and eminently worthwhile but it does not take us to the mainspring of historical technology; it does not throw much light on the actual processes whereby technology advances.

Secondly we have the efforts of the general historians and, in particular, of the economic historians. They have written a number of illuminating studies of the ways in which technology has contributed to the general progress of society and to the growth of wealth in particular. The economic historians have certainly studied the history of technology from a more general, or less parochial point of view, than the industrial archaeologists for they do not, at least in principle, attempt to limit their studies to Great Britain or to particular regions.

Thirdly, there are antiquarians, as we should term them, who without the romanticism of Smiles and his disciples or the philosophical approach of the economic historians take specific objects, or families of objects, and study them to the exclusion of all else: the thing in itself is sufficient, whether it be an old motor car, a tramcar, a locomotive or a clock.

While not in any way criticising these three approaches to the history of technology the writer wishes to suggest that

---

* There had been a number of Scottish historians of science before Whewell. It is only fair to add that none of them had reached his high standard of scholarship.

another and complementary approach may be no less valid even though it has not been developed to anything like the same extent. In particular it is possible to envisage a history of technology which is closely related to the history of science and to the history of ideas generally. Technology so regarded is not to be thought of as the dependent variable, drawing its ideas from and parasitic upon science; rather it is an equal partner contributing at least as much to the common stock as it draws out. It is, in fact, this approach that is suggested in this work; and it provides us with criteria for the determination of what is important and what is unimportant; what should be included and what left aside.

This does not imply that only the most scientific branches of technology merit consideration. Far from it; for one of our purposes is to enquire into the methods of technology and the modes of technological advance. Not all of these are seen to be scientific; some are indeed empirical. But once technological innovation has become commonplace and self-sustaining (like science) a rationale becomes possible which enables us to exercise discriminatory judgments. In essence it is a matter of the consequences and generality of a particular innovation together with the originality and insight displayed by the inventor.

There is nothing revolutionary about this approach. As the word 'technology' implies a relationship between science and the industrial arts it is, on the contrary, the natural one to choose. Why then, in spite of the works of men like the late Professor Wolf and, more recently Professor Forbes and the late Professor Dijksterhuis,* has it not been established? There appear to be several reasons. For one thing there has been the continuing lack of communication between the disciples of Smiles and those of Whewell; there is the intensive specialisation of studies which tends nowadays to inflict division rather than encourage unification: and, finally, there is the still relatively undeveloped state of the history of science.

Historians and philosophers of science have tended, until recently, to take the view that science is a 'pure' intellectual

---

* Cf. A. Wolf, *A History of Science, Technology and Philosophy in the XVIth and XVIIth Centuries;* and *in the XVIIIth Century.* R. J. Forbes and E. J. Dijksterhuis, *A History of Science and Technology* (Penguin Books), activity, having nothing to do with utilitarian objectives and concerned solely with the higher reaches of abstract thought. This attitude has been due in part to the positivist philosophies held by the majority, or at any rate the most vocal, of philosophers of science; in part it has been due to the laudable desire to protect science from those political movements that, in recent decades, have proclaimed science to be a characteristic activity of particular races, classes or ideologies. And it has owed something, regrettably enough, to the immemorial academic snobbery of philosophers.

Times however change. During the nineteen-thirties there was much interest in such problems as whether or not electrons exist, whether a rod travelling at nearly the velocity of light would really contract and what meaning could be ascribed to the statement that space-time is curved. But events since 1939 have provoked us into looking at science and technology afresh. How was it possible to organise scientists and technologists so that an atomic bomb could be made? What are the prospects for organised science and technology? What are the particular moral responsibilities of scientists and technologists? We can now see that society is deeply and irreversibly implicated in the affairs of technology and science; and we realise that science is socially determined as never before. This cannot be changed: science, technology and political affairs are now closely interwoven, the warp and weft being national defence, education, institutionalised research, the needs of science-based industries and a host of other sensitive issues. The problems here are innumerable; material for countless extended debates and learned treatises. But the impact of technology and science on society and—no less important—the impact of society on technology and science are themes that lie outside the scope of this book.

The following work is, then, an attempt to identify the turning points in the history of technology and to elucidate the principles that were involved. It is not concerned with the details of individual inventions; for accounts of these reference must be made to the specialised studies mentioned in the bibliographies at the end of each chapter. Thus, for example, there are simple, even elementary, descriptions of the medieval inventions of the weight-driven clock and the printing press for these things affected European thought—including technology and science—to an extent that few other inventions, including those of the present day, have done. But it is not necessary to go beyond the general principles; there are available several excellent specialised studies and these are mentioned in the bibliography.

On the other hand it is surely impossible to understand the development of modern technology without some understanding of Galileo's mechanics, the thermodynamics of Sadi Carnot, Kelvin and Clausius and the field theory of Faraday and Maxwell. The reader must judge the validity of this statement from the accounts given below of these three major scientific advances and of their 'two-way', cause-and-effect relations with technology. In the meantime it is worth noting that modern technology is still, to a great extent, built on the scientific work of the nineteenth and earlier centuries; work which has proved far richer than even nineteenth century scientists expected. By contrast, nuclear energy, a characteristic achievement of the present century, has still only had a marginal impact—apart, that is, from its sinister implications for the military arts.

Important technologies, such as the chemical, metallurgical and extractive and agricultural ones, have been excluded from the book simply for reasons of space. These technologies are, in any case, autonomous to a great extent, and merit individual treatment in separate volumes; so too does structural engineering about which very little is said in the following pages.

Finally I must thank my friends and colleagues, Dr Robert Fox (University of Lancaster), Dr R. L. Hills, Dr A. J. Pacey and Dr J. R. Ravetz (University of Leeds) for many stimulating discussions; some of their ideas are put forward in this work, although I hope always with due acknowledgment. I am grateful also to a number of postgraduate and undergraduate students: I owe a great deal to their interest and intelligence. Unfortunately they are too many to mention by name.

Chapter 1

The Background

Even a very simple human community can hardly function without a modicum of technics. Such a community must, if it is to rise above the level of the wandering food-gatherers, command the rudiments of agriculture; it must have invented or adopted simple agricultural implements; it must have acquired the basic skills needed to build houses, to weave cloth, to make weapons and to manufacture simple domestic utensils such as cooking pots, ovens, cutlery, querns and basic furniture.

A comparative study of different cultures suggests, however, that once a community has achieved a certain degree of control of its environment and the division of labour has appeared, then in effect it has to make a choice. Technical progress can be maintained or a form of hierarchical society established in which slavery of some sort or another provides the alternative to technics. The expendable slave competes with the expendable machine and, no doubt, in many cases the advantage lay, or appeared to lie, with slavery: machines are difficult to develop and, in the early stages, liable to frequent break-downs. On the other hand the defect of the slave-based society is that, quite apart from ethical considerations, it tends to be static and unadventurous. Slave labour is inefficient compared with free labour and unless an unlimited supply of slaves is assured any cost advantage must soon disappear. Lacking any objective understanding of the past— that is, lacking history—the hierarchical and slave owning societies of classical antiquity failed to appreciate the great progress that had been achieved by and through technics. Rather they seem to have interpreted the past in mythical terms, as a 'golden age', and to have diagnosed their present as decadent; the future they seem to have regarded with despair. Such an attitude amounted to a negation of technics.

A number of hypotheses, some of them very persuasive, can be put forward to explain the eventual stagnation, coupled with technical unproductiveness, of the later Graeco-Roman world. Here we merely note the fact and pass on to observe that the social circumstances favouring the development of technics and its evolution into *technology*, which we shall briefly define for the present as technics based on science, are exceptional and have occurred in only a few very successful cultures. Only in one—our own—has it been fully achieved.

Technics and technology are basic elements of civilisation. They constitute a common denominator, an essential factor, for without them civilisation would be impossible. And yet, in spite of this, the historical study of technics and technology is still at an elementary stage. In the nineteenth century, when the effects of industrial revolution became apparent, there was considerable interest in the subject—this is confirmed by the writings of Samuel Smiles—but thereafter it waned, the low point of indifference being touched at the beginning of the present century. Since then there has been a slow but steady revival, aided on the one hand by the activities of the New-comen Society (founded in 1919) and kindred bodies but retarded on the other hand by academic conservatism together with a curious but apparently deep-seated dislike of the whole topic on the part of certain writers. There has existed in every age that affected contempt for the 'banausic' arts that so corrupted Greek thinking and which may ultimately have made Greek (Hellenic) science so sterile. Technics are associated with lower-class activities and that means, in one word, slavery. But these old prejudices can hardly account for the low esteem in which technics and technology are held in some quarters even at the present day. Thus for example a recent article in *The Times* newspaper by Dr A. J. Toynbee was headed 'Christianity's Chance to Triumph over Technology', while a letter from Dr F. R. Leavis, in the same newspaper, criticised the 'meaninglessness of our . . . technologico-Bentha-mite world'. Dr Toynbee did not think it necessary to mention the points at which Christianity and technology might be in conflict and Dr Leavis did not indicate a rational procedure for determining the meaningfulness of this or any other world. But these points do not concern us: what is significant is that distinguished writers can be so hostile to technics and technology. How can we account for this not uncommon attitude? After all, some of the political philosophy of the past two hundred years has been much more subversive of public order, civilised behaviour and international peace than technics has ever been; a good deal of modern literature and drama has been hostile in spirit to democratic and humane sentiments; and if technics have been misused and misapplied in wartime, that surely is something for which we are all equally to blame? It is probable that deep seated dislike of technics is an atavism. The fear of nature runs deep, it seems. There is, as the French scholar Robert Lenoble remarked, a tabu on the natural. The scientist is bad enough—he pries into the dark, awful secrets of nature; but the technologist is worse—he tries to bend nature to his will and usually succeeds in doing so. Metaphorically one shudders at his impiety and longs for the comfort and emotional security of the ancestral cave. It is not 'natural'; we cannot and should not try to do what nature does. And this is a prejudice which is clearly recognised by those whose devotion to truth is not perhaps as strong as their devotion to other considerations. We are all familiar with the television advertiser who claims that a particular product 'has pure natural goodness . . .'. The operative word is 'natural'. It represents something that we cannot attain or aspire to: it is always outside our reach. The implicit ideal would appear to be that of natural products in a natural unspoiled world with a natural society of men and women. But there is no such thing as a 'natural' society or a benevolent 'natural' world and the value of individual things must be judged strictly on merit. Anopheles, the mamba, and the cholera bacterium are 'natural', while insulin, penicillin and the Salk vaccine are artificial.

There were no scientists in the ancient world, for science as a social institution did not exist and indeed did not begin to be significant until the seventeenth century. This is not to deny that a good deal of the groundwork of science was prepared by the ancients and that this was, in turn, to be of importance for the development of technics and technology. The Greeks, in particular, held the rational view that all natural phenomena are determined by laws or principles and not by the arbitrary whims of deities, spirits and demons. They enormously extended mathematics and invented the idea of proof. They put forward mathematical theories of the heavens and laid the foundation for a science of astronomy. Finally a small group of expatriate Greeks living in or near the city of Alexandria demonstrated a practical boldness in experiment coupled with a professional attitude which sharply distinguished them from all other ancient Greeks. These men made contributions to cartography, theoretical and practical mechanics, practical astronomy, chemistry and medicine. It is possible that this promising beginning was the outcome of a cross-fertilisation between the Greek and Egyptian cultures.

However, these things apart, Greece and Rome are no more interesting to the student of the history of technics than are such cultures as those of China and Islam. As regards the practical arts their greatest bequests to the western world were administrative and legal systems. The Romans in fact extended and maintained their empire by means of an efficient army and a highly competent, yet generally just and tolerant bureaucracy. It is perhaps worth remembering that none of the wonders of the ancient world were Roman achievements.

The break in technics between the ancient and modern worlds is almost complete and this, in part, accounts for the still common but quite erroneous belief that the Romans were first-class engineers and technologists while the medieval Europeans were incompetent in these fields. This, it is alleged, is confirmed by the fact that the Europeans let the Roman roads fall into disrepair and failed to build aqueducts, market places and villas with under-floor heating.

On the contrary it is easy to argue and, to the extent that proof is possible in such cases, to prove that the medieval period was one of the greatest periods of technical advance in history. Apart from the great cathedrals, abbeys and castles—surely not inferior to aqueducts and villas?—this was the age that produced a number of the basic inventions on which the whole secular fabric of our present civilisation, to say nothing of our technology and science, rests. In comparison with the technics of the medieval world AD 800 to about AD 1500 those of the Greeks and Romans fade into insignificance.

There are, from a general point of view, several interesting features about the development of medieval technics. In the first place it began with almost total dependence on imported ideas and inventions which commonly originated in China, India or among the Arabs. The latter performed the additional service of transmitting the inventions to Europe and the points of transfer were, naturally enough, those places where Islamic civilisation and European culture met: in Italy and southern Spain. The Arabs had inherited and extended Greek astronomy and mathematics as well as Greek philosophy but they added a significant technical component, missing from the original Greek achievement. They made contributions to medicine, especially ophthalmic optics, and they developed the study of chemistry as well as a wide range of chemical technics. They made original improvements to navigation and ship building, to metallurgy and architecture, to irrigation and water supply, to cosmetics and the culinary arts. The transmission of inventions, whether their own or other people's, to Europe certainly proves the excellence and liberality of early Islamic civilisation. But that the traffic was one-way does not prove the inferiority of the Europeans. Indeed it proves exactly the opposite, for a technologically progressive society is obviously one which is both willing and able to accept and apply inventions *from whatever source they may come*. This is surely as true of major communities, such as nations, as it is of much smaller ones, such as manufacturing concerns. There is no particular shame about being an 'adoptive' community if that is the only, or the quickest, or the best way to become technically progressive. Subsequent history indicated that this was what happened in the middle ages.

If the Europeans were imitative and adoptive in the early middle ages, the English were, in the seventeenth and early eighteenth centuries, commonly regarded by their European neighbours as unoriginal imitators of continental inventions; and this was just before the great efflorescence of English inventive genius that accompanied the industrial revolution. More recently the Japanese have had the same reputation as copiers; this time of the Europeans, English and Americans. But it is clear enough now that the Japanese have acquired their own technological momentum. In brief it seems probable that imitation, adoption and adaptation are essential steps whereby the art of invention is transmitted from one culture to another. But willingness to submit to instruction, as it were, must be there in the first place. An interesting instance of what happens when the essential humility is lacking was provided by China. The Chinese produced a remarkable number of inventions and they were either willing to allow or unable to prevent the spread of these ideas to other cultures. On the other hand the Chinese seem only rarely and in special instances to have been willing to allow the importation of foreign inventions. The consequence was that Chinese technics eventually languished.

We infer that in medieval Europe the town and village craftsmen—smiths, wheelwrights, millwrights, gold- and silver-smiths, masons and carpenters—steadily improved their skills and thereby their capacities to make new products so that before too long they were capable of making inventions of their own. The more numerous, the more skilled and independent and the more confident the craftsmen became the more likely they would be to launch out into new inventions. So much would seem to be almost self-evident; but we have no idea what basic drives initiated and thereafter maintained the—essentially social—mechanism of technical advance.

One of the more conspicuous features of medieval technics, at all levels, was the boldness of the innovators. In comparison, the ancients and, to be fair, the craftsmen of most other cultures, seem to have been a timorous bunch; unwilling for fear of the tabu of the natural to wrest the full benefits of command over nature that their knowledge and skills should have given them. Thus for example the practice of human dissection in pursuit of medical knowledge and as an aid to medical teaching was gradually sanctioned in the medieval universities while it remained forbidden in contemporary and earlier civilisations (ancient Egypt had been a notable exception). Chemistry, both in the form of the tabu-breaking alchemy and as a practical art—the making of paints and dyes, mordants and glass—was practised in medieval times while the great mining industry was rapidly developed. The significance of mining was that it was carried on by free men, not by doomed slaves. Again we detect a boldness missing from ancient technics: prying into the heart of the earth was not something the ancients cared much about. Finally we remark the extraordinarily courageous voyages made by Italian, Portuguese and Spanish navigators in the fifteenth century which culminated in the discovery of the world—one can hardly rate it less—by about 1520.

There was, then, a boldness and a lack of inhibition about both the medieval approach to nature and medieval technics. They were not frightened about meddling with nature by carrying out chemical operations, dissecting dead bodies or voyaging across the oceans to new continents. Traditionally, legends warn men of the awful perils of violating the tabu on the natural: the punishment inflicted on Prometheus for trying to do what only the gods can accomplish; the disaster of Daedelus and Icarus who tried to fly like birds, are just two of the many salutary warnings against impious violations of nature. But the medievals seem to have turned their backs, almost scornfully, on these inhibitory legends. In the case of endeavours to make perpetual motion machines or to transmute base metals into gold the challenge to ancient prejudice as well as to the authority of Aristotle was direct and uncompromising; in the case of the navigations, dissection, practical chemistry and mining the challenge was no less direct and ultimately much more fruitful.

If we seek the reasons for this remarkable emancipation of the human spirit we shall find it difficult to avoid the conclusion that the basic cause must have been the universal religion of Europe. Christianity laid positive duties on men while at the same time rejecting the enervating fatalism of so many ancient and Eastern religions. It provided the right spiritual armour for a man confronting nature in the raw. There is one sovereign Deity who is the ultimate protector and saviour. One does not need to—indeed one must not—appeal to a host of minor deities, godlings, spirits, ghosts. Of course the varieties of Christian religious experience are very wide and the number of philosophies which, in their time, have been reconciled with Christianity considerable. We cannot therefore argue that Christianity was always and everywhere uniform in effect. But we do suggest that a central component of its doctrines had the effect we have just described and thus emboldened medieval man to develop his technics and later, in fact, to establish his science.

Another and not less important effect of Christianity was the social structures to which it gave rise and the values that their characteristic institutions induced. A universal monotheistic religion, dogmatic in its administration but, thanks to the revival of Greek and Roman learning, rational in its interpretations must have given Europe a degree of intellectual coherence never before matched over such a wide area and embracing so many peoples (unless perhaps China be the exception). This unity of Christendom was reflected in the inauguration of those characteristic medieval institutions: the universities. In these corporate bodies the dominant philosophy might change from neo-Platonist to Thomist Aristoteleanism but the essential universality coupled with a syllabus based on the seven liberal arts and crowned by the three major faculties of divinity, law and medicine meant that learned men in Salerno, Bologna, Paris, Wittenberg and Oxford spoke the same sort of language and thought in the same categories. The consequences of this unification of rational thought through the great communications system of the universities were profound indeed. In fact the present international communications network of technology and science is a direct, if very much more elaborate, descendent of the medieval system.

If the curse of slavery was removed from medieval society, or at least was mitigated into the milder institutions of serfdom and vassalage, then the stimuli towards technical innovation were so much greater. Usury might be frowned on but imperfect economic incentives may have been counterbalanced by certain ideological incentives to innovation that were even more effective than the immediate prospect of gain, at any rate in the long run. For while it is true that a utilitarian bias in society must operate in favour of innovation, it is none the less also true that some of the most important inventions have been made for reasons, or under circumstances that transcend utility and can perhaps only be called ideological or idealistic. In this respect the middle ages may have achieved a happy balance in its incentives for invention, uniting immediate utility with an idealistic component which encouraged inventions that could pay off in the long term. So effective were these various incentives, acting at different levels, that medieval technics was, as we may say, *precocious*: in advance of the somewhat limited scientific knowledge of the time. It is fair to add that, taking the long view, sustained technical progress began some hundreds of years before the scientific revolution of the seventeenth century. But before we go any further let us look at the spread of technics in Europe and enumerate the major inventions and innovations of the period AD 800 to AD 1500 the period commonly called the middle ages. We shall then go on to discuss two major inventions of the period and in this way clarify and justify the distinction we have drawn between inventions of immediate utility and those whose values become apparent only in future years. In this way we shall be able better to appreciate the range and vitality of medieval technics but also gain some insight into the nature of the process of innovation in all ages.

A map of the distribution of medieval technics is at once revealing. It indicates clearly enough the heavy concentration in Notrhern Italy and Southern Germany: the region of alpine technology as we may call it. German medieval technics had a definite bias towards mining and metallurgy and this, it was much later observed, probably accounted for German excellence in the closely related science of chemistry; an excellence that has lasted up to the present day. On the other hand Italian technics was associated with art, architecture and mechanics and it is worth remarking that modern Italy has a high reputation in such fields of precision engineering as high quality automobiles, typewriters and scientific instruments. By comparison France was at that time backward and England very backward. This may have been related to the fact that while such Italian universities as Padua and Bologna were strongly utilitarian in their academic bias, Paris and Oxford were much more closely associated with philosophy and theology. But whether these academic interests were causes or effects is altogether another problem.

Medieval technics was quite unlike those of Rome, Egypt and Babylon, which were grandiose and extravagant. There was no Great Pyramid, no Colossus, no Colosseum (the very name suggests megalomania), instead all over medieval Europe craftsmen were experimenting with new methods, new machines, new devices to save labour and improve the quality of life. Properly regarded, the technics of the middle ages was

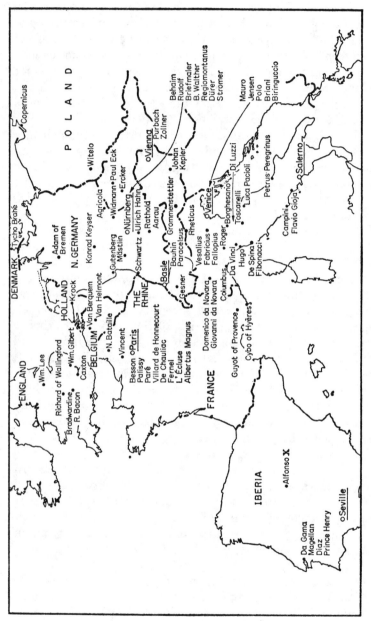

Map. *The distribution of medieval technics.*

vastly more impressive than the conspicuous consumption, which masked the bankruptcy of invention, of the ancient empires.

In agriculture, the basic industry, we find that the stirrup, the rotation of crops and the Saxon wheeled plough, that enabled heavy, fertile soils to be cultivated, all came into use in western Europe before AD 1000. Nailed horse shoes had been introduced and so had the 'modern' horse collar, that allowed the animal to exert its maximum pull with its shoulders; older collars, round the animal's neck, had obviously been much less effective. The windmill, unknown in the ancient world, was invented and by the time of the Norman conquest some 6,000 water-wheels were in use in England. In the following centuries the applications of water power diversified to a variety of processes: sawing wood, forging iron and fulling as well as grinding corn. Over the same period the crank was introduced and improvements were made in methods of transmitting power. The manufacture of gunpowder—a Chinese invention—enabled firearms to be invented, the first cannon being recorded as early as 1318. The blast furnace appeared, metallurgy was developed and methods of casting bronze and iron were improved. The lodestone was made the object of a theoretical study in 1269 and the mariner's compass came into use. The manufacture of paper was introduced (another Chinese invention) and spectacles were invented towards the end of the thirteenth century. A variety of new chemical substances were adopted, due almost entirely to the Arabs. These included camphor, calomel, tinctures, dyes, pigments, mordants and medicaments. The study of mathematics began to revive and the excellent Arabic system of numeration replaced the clumsy Roman method. These, in short, represent only a few of the major innovations of the period; innovations that ensured that by the end of the middle ages the technical standards of western Europe were higher than those of any previous civilisation.

An interesting example of medieval inventiveness which was to have very fruitful consequences in the eighteenth century was the invention of the spinning wheel. A large, but lightweight, wheel free to rotate on a horizontal axle, drives a small spindle by means of an endless cord. A short length of yarn is drawn out from a mass of cotton or wool held in one hand and is twisted between the 'spinster's' fingers to give a strong thread. The thread is then wound round the spindle and the big wheel is set turning with the free hand. By drawing the mass of cotton or wool away from the spindle and paying it out between the fingers further thread can be spun, rapidly and easily, by the turning of the spindle. When the arm's length of thread has been spun in this way it can be wound on to the spindle merely by swinging it through about 90° and continuing to rotate the spindle rapidly. By this means the 'productivity' of spinning was greatly increased.

There can be little doubt, however, that of all the great medieval inventions none surpassed the weight-driven clock and the printing press, measured by the scales of inventive

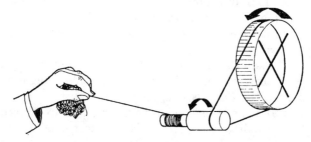

Figure 1.   *The action of the spinning wheel.*

insight on the one hand and social, philosophical, even spiritual, importance on the other. Indeed these inventions were so fundamental that they still condition our lives in ways and to an extent that few other inventions have done, right up to those of the present day. The clock and the printing press are, in fact, the twin pillars of our civilisation and modern, organised society is unthinkable without them. We shall therefore conclude this introduction with accounts of these two inventions. We shall find too, that they can tell us a good deal about the nature of the inventive process itself.

THE WEIGHT-DRIVEN CLOCK

The position of the sun in the sky is the ultimate determinant of time; and one of the simplest, if not also one of the earliest, methods of finding the time was to note the change in position of the shadow cast by a fixed object in daylight. But while sundials might be effective in Mediterranean lands they would be much less useful in northern countries where the sun is so often obscured by clouds. In the north, therefore, an alternative method of time-keeping was desirable; and the possibilities were either an hour-glass, or a calibrated candle, or a waterclock. However the first two were imprecise instruments, suitable only for measuring short intervals of time. The waterclock was more promising but the difficulty of ensuring a uniform flow of water proved formidable. An added, if not so serious, complication was the common practice of dividing the period of daylight into an equal number of hours. This was reasonable at a time when all communal activities had to cease at nightfall because there was no adequate means of artificial illumination. But it meant that, as the length of day— and thus of the hour—varied with the season, the sun-dial or water-clock had to be calibrated with a complex set of hour marks so engraved that the 'correct' time could be read at any time of the year.

The first appearance of the weight-driven clock was as mysterious as it was intriguing. The first firm indication of such a machine was in the year 1286, but who made it and for what purpose is unknown. It may well have been an astronomical clock and descended therefore from a line of models or instruments intended to imitate the motions of the heavenly spheres and the planets they carry. Its design may also have reflected the speculations of some millwright, accustomed to fabricating gearing, knowledgeable about uniform motion and gifted with an astonishing insight into mechanical principles; it seems likely, too, that the skills of an able blacksmith may have been called in.

The difficulty about driving a clock by means of a falling weight is simply that, as Aristotle knew, the weight accelerates as it falls so that the clock would tend to run faster and faster. One can, of course, retard the fall by means of a brake so that when the weight reaches its terminal velocity it falls, like a ballbearing in treacle, with uniform velocity. But the trouble is that as the brake and the rope wear smooth so the terminal velocity will gradually increase.

Such a solution is therefore unsatisfactory and we are left completely in the dark about the steps by which some unknown genius—or geniuses—invented the escapement mechanism which solves the problem and constituted perhaps the greatest single human invention since the appearance of the wheel. The principle of the escapement—the heart of the invention—can best be explained with the aid of a simplified diagram. The wheel (Figure 2) is fitted with short horizontal pegs uniformly

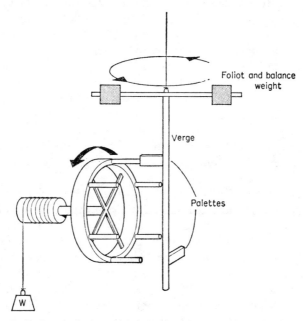

Figure 2.　*The mechanism of the simplest form of weight-driven clock.*

spaced round its circumference and is driven by means of a falling weight at the end of a cord which is wound round and round the axle. Just in front of the wheel is a vertical rod—the 'verge'—to which are fixed two small plates—'palettes'—slightly more than 90° apart. These palettes are placed so that they mesh with the pegs on the wheel. The verge is suspended freely by a length of rope or cord and at the top it carries a short horizontal rod—the 'foliot'—which has two balance weights at each end: the position of these balance weights can be adjusted.

The operation of this mechanism is quite simple. Let us suppose that the wheel is moving in the direction of the arrow. When the first peg strikes a palette the motion of the wheel is at once checked by the inertia (we shall use modern terms) of the verge, foliot and balance weights. The driving weight then slowly accelerates this system until the peg has pushed the palette out of the way and the driving weight can, for a moment, fall freely. The turning of the verge has, however, brought the lower palette between the pegs of the wheel so that the next lower peg, moving of course in the opposite direction to the ones on top, almost immediately strikes the lower palette. The swinging motion of the balance weights is checked and the fall of the driving weight slowed down again. Once more the latter exerts an accelerating force on the palette and the system to which it is attached, but this time in the opposite direction so that the first, upper palette is brought back again into mesh with the pegs of the wheel. The process then repeats itself indefinitely with the foliot and balance weights swinging to and fro and the falling weight repeatedly checked by being compelled regularly to reverse the motion of the balance weights and their suspension. By judiciously arranging the proportions of the balance weights to the driving weight and the distances of the driving weight from the verge the whole subtle machine can be made to move with a regularly inter-rupted motion, the fall of the driving weight being periodically checked and re-started so that the average or overall motion is uniform.

Now if we pay attention to the principal features of this invention two things strike us at once. Firstly, there is the brilliant mechanical, or kinematic, insight that the motions in opposite directions of the upper and lower parts of the escapement wheel can, by means of palettes at 90° to each other, be converted into the cyclic to-and-fro motion of the foliot and balance weights. In brief, this implies inventive genius of the highest order.

Secondly, and hardly less impressive, there is the mastery of the dynamical principle whereby the fall of the driving weight is *uniformly* checked by being required to impart accelerative motion to the balance weights. Any understanding of the theory behind this principle was a long way in the future: not until the times of Galileo and Newton could it be scientifically accounted for. We are therefore justified in describing this invention as 'precocious' for, like other medieval inventions, the principles of its operation were far ahead of contemporary scientific understanding. On the other hand if we try to trace the antecedents of the mechanical clock we immediately run up against a blank wall for very little information is available. All we can suppose is that there must have been a protracted series of attempts to devise mechanical models to reproduce the motions of the heavens and that finally these machines issued in the first weight-driven clocks, the (hour) hand of which showed the position of the main planet—the sun—in the course of his diurnal journey round the earth. Certainly many of the first clocks were astronomical ones and some were of such elaborate design that they could indicate the positions of the sun, the moon, the other five planets and even the motions of the tide.

Clocks based on these principles were erected in the late thirteenth and early fourteenth centuries, being placed on cathedrals, churches and castles. Early pioneers of the art of clockmaking were Richard of Wallingford and the de Dondi family and after their work development became systematic. The balance wheel and 'clockwork' (spring) drive were introduced about 1500 and the watch was invented. But the substitution of spring drive for gravity drive posed a new problem. Coiled springs, unlike gravity, tend to run down and exert less and less driving force as they unwind. To compensate for this a method of storing up energy—as it would now be termed—at the beginning and then releasing it as the spring unwound was devised. This was called the 'stackfreed' and it worked by causing a rotating cam to drive out a pivoted arm against the pressure of its own elasticity until, as the cam rotated further, the process changed and the energy previously stored in the arm was progressively released to supplement the waning force of the driving spring. Another device to achieve the same end was the fusee, placed between the mainspring and the hands. A cone and a cylinder, parallel to each other, were coupled by a cord so that as the spring unwound the gear ratio progressively changed in such a way that less and less force was required to turn the shaft. These two inventions are illustrated in Figure 3.

Later still came Galileo's invention of the pendulum clock, using the isochronous property of the pendulum instead of the inertia of the balance weights, together with such improvements as the introduction of the dead-beat and anchor escapements. The apotheosis of the mechanical clock came in the eighteenth century with Harrison's marine chronometer.

We may assume that, as has been suggested, the mechanical clock was originally an astronomical device: the circular

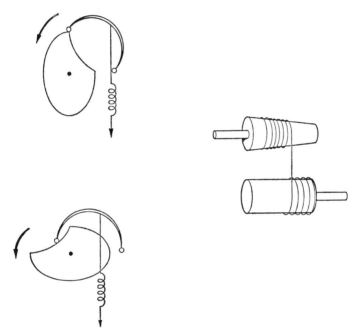

Figure 3.   *The principles of the stackfreed* (left) *and the fusee* (right).

clock face representing the circle of the heavens and the hour hand indicating the sun's position on that circle. As the motion of the sun is uniform and the same in summer and winter it would have been absurd to have calibrated clock faces with different hour marks, appropriate to the different seasons. It is hardly surprising therefore that with the spread of the weight-driven clock the fixed, uniform hour—the astronomer's measure—replaced the old, variable one. More, the intro-duction of the mechanical clock seems to have changed fundamentally men's ideas about time for it could now be conceived of as flowing on ineluctably and independent of particular events; in Newton's words, it could be thought of as absolute, true and mathematical time. No longer was it merely the interval or period between different events; it was a steady, unchanging progression. By the fourteenth century the art of the clockmaker had advanced sufficiently for the hour to be divided into sixty minutes and the minute into sixty seconds. Since those early days the remorseless tick of the escapement-controlled clock has marked the unending passage of time: the dimension against which all the affairs of men are ordered.

Astronomers together with philosophers and theologians may well have had an awareness of the steadily flowing stream of time—the moving image of eternity—but the invention of the mechanical clock made it explicit and obvious to all men. Furthermore, without a precise, easily available and generally accepted time standard the administrative, commercial and industrial arrangements of advanced civilisations would be impossible. Although we now have such sophisticated time measuring devices as crystal oscillators and atomic clocks, we still, in the last resort, arrange our affairs in the framework set out by the medieval invention of the weight-driven clock.

The development of the art of clock making gave rise to a new and very superior class of craftsmen. These men became skilled in the design and manufacture of gearing and in the different methods of transforming and applying motion. Inevitably, it seems, they transcended the confines of their craft and before long we find them being employed to design and superintend the construction of waterwheels; later still in the period of the British industrial revolution, clock and watchmakers figured as key engineers in the construction and operation of textile machinery. There was, therefore, a direct link between the medieval invention of the mechanical clock and the great industrial transformation that began in eighteenth-century England; a transformation that was to indicate the way to public affluence and now holds out some hope for the eventual ending of world poverty and famine.

In sum, the mechanical clock altered men's understanding of their world, changing their philosophy of time and in this way making its contribution to the great scientific revolution of the seventeenth century. So too, on a different plane, did the imagery of clockwork; the system of the heavens conceived as a great piece of clockwork with God as the celestial clockmaker proved very acceptable to the seventeenth- and eighteenth-century scientists. Correlatively, clockmaking was for long the pinnacle of mechanical arts and the training ground as well as the inspiration for practitioners in other branches of mechanics.

Generally, the mechanical contrivances of the middle ages and the renaissance tended to be over-complex, over-elaborate as if man delighted in their new skills and had little regard for economy of operation and mechanical efficiency. This attitude was to change very radically in the seventeenth century but, in the meantime, it was hardly surprising if mechanics and magic came to be somewhat confused and the distinct talents of the mechanic and the inventor ascribed to magical practices. Machines, after all, were commonly regarded as devices for cheating nature and the machine maker would therefore have been regarded as a man of altogether superior knowledge and powers. Among the unsophisticated the legend of the mechanic-magician lasted a long time; it was perpetuated, for example, in the figures of Spallanzani and Dr Coppelius, the doll-maker.

THE PRINTING-PRESS

The second great medieval invention that we shall discuss is that of printing by means of moveable type cast in an adjustable type-mould. This innovation is doubly interesting, for not only does it exemplify several features that seem to be common to most, if not all, inventions but it was also the first invention about whose antecedents, genesis and immediate consequences we know a good deal; not least we know the name of the inventor.

In a sense printing has existed from time immemorial. The Royal Seal, the punches used by gold- and silversmiths to impress hall-marks, the signet ring and, more recently, the bureaucrat's rubber stamp, are all simple forms of printing. But, before the printing of books, journals and newspapers became possible, several vital ancillary innovations had to be made. For one thing, a cheap and convenient raw material—paper—had to be available, printing ink had to be developed, the principle of the press had to be adapted (probably from the ancient art of wine making); but, above all, the problem of making type cheaply and accurately had to be solved.

It is reasonable to suppose that the demand for books was increasing during this period of technical development and geographical discovery. And furthermore that this demand was being frustrated by the shortage and incompetence of scribes and copiers. We know that by about 1400 such things as playing-cards and simple religious pictures depicting saints were being printed by means of stamps in northern Italy. But the cost of preparing a really elaborate stamp containing enough words and punctuation marks to constitute a page of (say) the Bible would clearly have been prohibitive. If to reduce this cost one cut down the number of words per page then proportionately more stamps would be required and one would be no better off, financially speaking. The impasse was complete and the problem must have seemed insurmountable to those who were concerned with book production. And yet the difficulty was resolved, and in solving it the middle ages left us with one of its greatest gifts: the printing press.

Although there are a few other claimants we shall probably not go far wrong if we accept the view of most scholars that Johan Gutenberg was the inventor of the printing press. He certainly produced printed books as early as anyone else and there are other facts in the case that make his claim very plausible. Gutenberg (1394 or 1399–1467) was born in Mainz where his father was goldsmith to the archbishop, a post of some responsibility and, one supposes, profit and respectability. Johan followed in his father's footsteps and became a skilled metallurgical craftsman; an accomplishment that, as events showed, would be essential for an inventor of the printing press.

Gutenberg's solution to the problem falls into two quite distinct stages. In the first place he had to solve it in principle, theoretically as it were, and then, secondly, he had to make his invention-in-principle actually work; in other words he had to take it through what today would be called the process of development. The first, or theoretical, stage was very simple. Instead of trying to carve out stamps for a whole page of print, or even for a whole word, Gutenberg proposed to make stamps for each individual letter, punctuation mark or other symbol. These small units of type are made in large numbers and each one is placed in an appropriate box: one box for capital 'A', another for small 'a', another for capital 'B' and so on. When he wants to set up a page of type he has merely to select the letters necessary to make the sequence of words, set them in a frame and, when the page is complete, clamp them tightly together. He has now only to ink the type face by means of an ink ball and press it against a sheet of paper to print his first page. After he has run off as many prints as he requires he unclamps and dismantles the type, returns each unit to its box and proceeds to set up the type for the next page of the book.

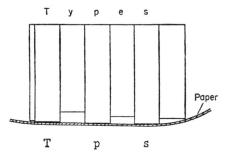

Figure 4. *Printing defects due to variations in the height of type.*

So much for the solution-in-principle; in practice Gutenberg must very soon have run up against two serious and related problems that must have cost him a good deal of time, patience, money—and genius. To illustrate these problems let us suppose that the small units of type are clamped together as shown in Figure 4. Some units, being slightly longer than average, project a little while others are somewhat retracted. Owing to these accidental variations in length some letters will be heavily printed while others will be faint or completely missed, being masked by their immediate neighbours. In fact, for effective printing all the units of type must have the same length; or, more precisely, the permissible variation is extremely small. But length is only one of the three dimensions and, as it happens, one of the other two is also critical. Let us now consider what will happen if the thickness of the type

B

varies beyond what is permissible. It turns out (Fig. 5) that the error is cumulative, getting worse line by line, until after a few lines the print becomes quite incoherent, letters from one line being juxtaposed with those of another.

The important point is that these snags could hardly have presented themselves to Gutenberg before he started out on his invention. They must have arisen during the second part, the development process. It was of course quite true that by using the greatest care and by throwing away all those units of type that did not reach the required extremely high standard of precision in the two critical dimensions Gutenberg might have been able to build up a stock of acceptable units of type. But

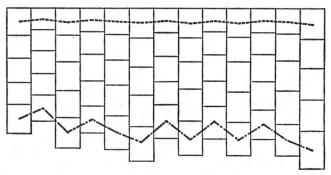

Figure 5.   *Printing defects due to variations in the thickness of type.*

the cost would have been enormous and all the advantages of the invention lost. Here then was a cardinal dilemma.

We do not know how Gutenberg solved it; what, that is, train of thought and experiment led to the ultimate answer, but in the event the solution-in-practice is no less brilliant than the original conception, the solution-in-principle. Complete accuracy could be obtained if all the type were cast in the same mould. Thus the necessity for extreme precision was transferred from the individual unit of type to the common mould in which they were all to be cast.

As finally produced by Gutenberg the type mould is adjustable, consisting of two L-shaped members, free to slide over each other (Figure 6). There were two good reasons for making the mould like this. In the first place, if the two sections can come apart the cast unit of type can be very easily removed and, in the second place, if the mould is adjustable one can vary the width of the units of type so that a small 'i' for example, takes up less space than a capital 'W'. In this way a more pleasing and legible print is obtained.

The bottom of the type mould is closed by a third member, the matrix (Figure 7), which is made of a soft metal like copper

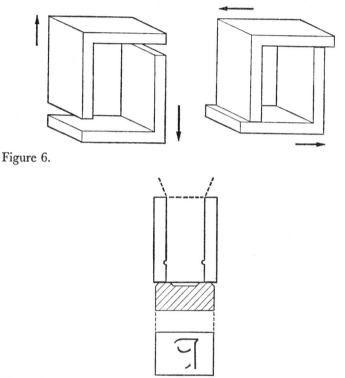

Figure 6.

Figure 7.

and which bears the impress of the particular letter which is to be cast. This letter has previously been punched in the soft metal by means of a hardened steel punch and, when enough units of type have been cast, a new matrix is substituted bearing a different letter and the process is repeated. The type metal used must not, of course, stick to the sides of the mould or to the matrix and for this purpose an alloy of tin, zinc and lead was found to be suitable. We can now appreciate how necessary it was for the inventor of the printing press to have had considerable experience in metallurgy and how Gutenberg's training must have prepared him for his work.

The completed unit of type, removed from the mould, is shown in Figure 8. The length of the type is fixed by a groove and a corresponding ridge, imposed by the mould, and which therefore serve to position all the units of type by causing them to lock together in the frame. The remaining problems were simple: the best composition for printer's ink was found empirically to be a mixture of lamp black and linseed oil and this was so successful that it was used up to the nineteenth century. The ink ball was made of leather and the principle of the screw press taken over, as we remarked, from the art of wine pressing.

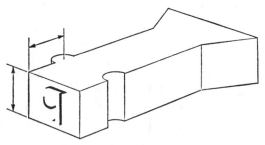

Figure 8.

Of its nature the printing press revolutionised the publication of books. It has been estimated that more books were made during the fifty years following Gutenberg (up that is to the beginning of the sixteenth century) than had been produced in the previous thousand years. A twenty-fold increase in productivity is something to reckon with and it would not be unreasonable to regard Gutenberg as the very first production engineer. But it was to be a long time before another invention as dramatic and important as his was to be made.

It is often remarked that the invention of printing was made independently and possibly somewhat earlier in China and in Korea. In the case of China the very nature of the elaborate script made it impossible that Chinese printing could be of the same form as Gutenberg's. It seems likely that porcelain casts were made of each Chinese ideogram as it was required; this was the 'rubber stamp' solution to the problem that would have been impracticable for the Roman alphabet. The Koreans on the other hand are said to have invented printing using cast metal type during the course of the fifteenth century. But it would be unwise to assume from this that the invention spread from Korea half-way round the world to Germany and was there copied by Gutenberg and his associates. On the contrary, the advanced state of German metallurgy, the immense though latent demand for books and the great skill that medieval European inventors and mechanics had acquired all combined to indicate that the time was ripe for the invention to be made independently in Europe. There is nothing at all implausible in simultaneous and independent inventions. When two or more communities have reached more or less the same standards of technical skill then the combination of wants, opportunities and stimuli provided by commerce, industry, administration and learning, makes it increasingly likely that the same things will be simultaneously and independently invented in each community. The more technically advanced communities become and the more communications improve, the greater the likelihood of simultaneous invention. It may now be said to be the rule rather than the exception.*

The mechanical clock and the printing press still determine the course of daily life to a quite remarkable extent and in this respect comparatively few of the individual inventions made subsequently can rival them. But they were only two out of the many innovations made in the medieval period the motives for which ranged from the economic through the humanitarian— increasing the quality of life by inventing such aids as spectacles —to delight in invention for its own sake or for ends which were mystical or magical.

Among the latter were such inventions as automata and the long sustained attempts to devise a perpetual motion machine. The last had complex roots. On the one hand there was the attempt to counter Aristotle's plausible assertion that motion, of its nature, demands the continued exercise of force whence it followed that perpetual motion was impossible. Only the

---

* Thus radar was invented simultaneously and independently in Germany, Holland, Britain and the U.S.A.; the gas turbine jet engine was invented simultaneously in Germany and Britain.

heavens were characterised by perpetual motion, and by definition, the gulf between heaven and earth was complete and unbridgeable. This challenge was too strong to be resisted by the heretical scientists, the anti-Aristoteleans of the middle ages. True, experience would seem to bear out Aristotelean mechanics in confirming that perpetual motion was impossible but surely a more exalted mechanics might be discovered, and on this higher plane it might be possible? The endeavour was doomed to failure but it bore curious fruit in that it made an indirect contribution to the subsequent development of clock-work: an early instance of the 'fall-out' principle.

From another point of view the idea of perpetual motion was by no means absurd; on the contrary the phenomenon was manifest daily. Apart from the eternal rotation of the heavens, the tides were in ceaseless motion, linked mysteriously to the motion of the moon, streams and rivers flowed on endlessly and the winds of the earth, although erratic showed no signs of dying away to a sustained calm. Every windmill, every water-wheel or tidemill was therefore a perpetual-motion machine of sorts; and if one could somehow master the principles of river, tide and wind one might hope to solve the problem completely and construct a simple self-contained perpetual-motion machine. The lodestone held out particular promise. It was thought, wrongly, to point towards the pole of the heavens and as the heavens are in perpetual motion the lodestone might offer a method of transferring this motion, or its principle, to the earth. But once again hopes were disappointed.

Magical or mystical ideas usually lie behind many of those medieval projects that, at first sight, appear to be remarkable foreshadowings of modern inventions. Roger Bacon, for instance, was for a long time honoured as a man who predicted things like trains, motor cars and aeroplanes, but there can be no doubt that the motive agencies he had in mind had less to do with the laws of thermodynamics or of electromagnetism than with the occult forces of the cosmos and the modes of application of power that he envisaged were less a matter of kinematics than of magical or transcendental mechanics.

However the dreamers and speculators can perhaps be most fairly represented as being on one wing of a large-scale movement, the centre of which was peopled by severely practical inventors like Gutenberg and his many less well-known contemporaries. On the other wing we find the artists and architects: indeed the early renaissance discovery of perspective and the work of architects like Leo Battista Alberti (1404–1472) meant that geometry was being put to the service of the practical arts of painting and building: in other words, mathematics was being applied to the understanding and control of nature. Dr A. C. Crombie has expressed the point very clearly:

> From the end of the fourteenth century it was the superior group that Leonardo Olschki called the 'artist-engineers' that took over from the academic philosophers as the pace makers of intellectual life. They were essentially a product of Italian urban society. Their achievement was to add to the logical control of argument of the philosophers a rational control of materials of many different kinds in painting, sculpture, architecture, engineering, canal building, fortification, gunnery, music.

This generalisation is equally applicable to Gutenberg's Germany if we modify the lists of technologies, placing rather more emphasis on things like mining, metallurgy and related chemical arts. Command over the materials of nature coupled with some intuitive insights into the principles that govern their behaviour, the achievements in short of medieval and renaissance technics, were to be key factors in the scientific revolution of the seventeenth century. But the achievements of the 'Alpine Region'—Northern Italy and Southern Germany—were not the only technical and quasi-technical factors that were to be important in the great scientific movement. There was also the quite remarkable phenomenon of the discoveries: a truly unique episode in world history and one that shifts our attention from the centre of Europe to the coasts of Spain and Portugal.

At the beginning of the fifteenth century knowledge of the world outside Europe was still very limited. Early travellers by land, like Carpini, Marco Polo and Odoric, brought back information about Mongolia and China; seawards and in the other direction the Norsemen sailed beyond Iceland to Greenland, and in all probability reached and even made settlements on the mainland of America (Labrador). Much further south other voyagers re-discovered the Canary Islands in 1336 and, later on, Madeira, the Azores and the Cape Verde islands. The rest of the world was a matter of conjecture, fable and the hypotheses of Ptolemy's *Geography*. However, the dramatic sequence of discoveries that were to follow these tentative probings were to change the picture completely, filling in the enormous blanks on the map and establishing the true form of the world. No doubt there were economic motives for these discoveries: the opening up of trade routes and the avoidance of the political difficulties of trading over land. But they also owed a good deal to what one can only describe as the new philosophy of nature; a down-to-earth and experimental attitude coupled with an unquenchable desire to find out at any cost. And the whole business was made possible by improvements in the arts of shipbuilding and of navigation—the development of the moveable rudder in place of the steering oar, for example, and the invention of the mariner's compass.

In the middle of the fifteenth century the Portuguese prince, Henry the Navigator, established a school of navigation—arguably the first technical college in the world—and encouraged his sailors to push further and further down the coast of Africa and further out into the unknown Atlantic. In 1455 the constellation of the Southern Cross was discovered and in 1481 the navigators crossed the equator and saw for the first time stars wheeling round a pole in the southern heavens. In 1492 the Genoese sailor Columbus reached the Americas; two years later Bartholomew Diaz reached the Cape of Good Hope. In 1497 Vasco da Gama rounded the Cape and sailed on to reach India, thus incidentally disposing of Ptolemy's supposed Southern Continent. Finally, in 1517 Magellan set out on the voyage that was to end three years later when one of his ships got back to Europe having circumnavigated the globe. Thus in much less than a century the Europeans had burst out of the confines of their little continent as known from time immemorial, had discovered new continents and oceans—including the largest of all, the Pacific—and had gone right round the world. They had touched on the fringe of the Antarctic continent and at long last knew beyond doubt the form and nature of their home, the earth. The laws of nature were everywhere uniform and the great sphere of the heavens, with previously unknown stars and constellations in the southern half, completely surrounded the small globe of the earth.

The social and political, economic and cultural consequences of the discoveries are too vast for us to attempt even to enumerate here. We content ourselves with noting that the unambiguous establishment of the true form of the earth made it easy for men to conceive of it in realistic and not abstract terms as a globe which, moreover, could be represented as such by means of a small model. And so, from about 1490 we find Martin Behaim, the cartographer of Nürnberg, making small model earths, or terrestrial globes, on which the progress of the discoveries could be marked. Two of the earth's near neighbours are globes—the sun and the moon—and as the earth, too, is established and accepted as a globe, it was now reasonable to take the next step: to propose that, if astronomical considerations require it, the globular earth might move, orbiting round the sun. Astronomical considerations did require such an hypothesis and, accordingly, the Copernican theory was enunciated in 1543.

The years that followed the great achievements of Gutenberg and Columbus seem by contrast to have been quiet and even unexciting; almost as if it were a time of assimilation and assessment. But in the middle and in the second half of the sixteenth century a series of books were published that were to stand as the landmarks of the new learning and the new knowledge. These were Copernicus' *De revolutionibus orbium celestium* (Nürnberg 1543); Andreas Vesalius' *De humani corporis fabrica*, (Basle 1543); Agricola's *De re metallica*, (Basle 1556); Vannocio Biringuccio's *De la pirotechnia* (Venice 1540); Lazarus Ercker's *On Ores and Assaying* (Prague 1574); Agostino Ramelli's *Le diverse et artificiose machine* (Paris 1588); Jacques Besson's *Theatrum instrumentorum et machinarum* (Lyons 1569). The pace of minor invention did not slacken, but overseas the Conquistador with his cannon and arquebus, military discipline and cavalry took the place of the navigators and set about establishing vast new empires on the ruins of the less successful cultures. It may be said that the New World and the New Age had both begun at the same time.

Chapter 2

The Seventeenth Century

Francis Bacon (1561–1626) is interesting from several points of view. For one thing he was that rare individual, an Englishman who made an impact on the international scene as a student of science and technology. For another he was the author of a famous scientific methodology: the so-called reformed inductive method which was for a long time greatly admired (at a distance) but which has recently been severely criticised (from slightly closer quarters). But perhaps Bacon's real importance was that he was the first man to lay down a social and political programme for science and technology. He was, in fact, a sociologist of science and technology.

There were good reasons for this. A notable lawyer and a man of affairs, whose whole career was spent close to the world of action, he was well placed, if inclined, to assess the significance of the new currents of thought and invention set loose by the achievements of the previous centuries. As it happened, Bacon was so inclined and the result was a series of prose works whose combined effect was as important as it is difficult to assess. He was certainly correct in his opinion that Aristotle's methods were scientifically fruitless; but his own reformed inductive method was equally open to objection. In essence the procedure he recommended was that of a legal tribunal. The philosopher investigating nature must, like a good judge, begin with an entirely open mind. He must take evidence from one side and then from the other before he attempts to sum up. He delivers a verdict on the strength of the evidence and this verdict constitutes scientific discovery. It was in this way that Bacon announced the conclusion that the Form of Heat (the mode of thought is still Aristotelean) was Motion.

But we are not concerned with logical objections to Baconian inductivism. We are more interested in his views as would-be planner and administrator of science and technology.* His heart is plainly in the right place for he observes (in *Novum Organum*) that: 'The true and lawful goal of the sciences is none other than that human life be endowed with new discoveries and powers.'

His opinion of the Greeks was not very high. He dismissed their knowledge as verbal and barren: they had no history to speak of, only fables and legends of antiquity; they knew only a small part of the world, being ignorant of Africa south of Ethiopia, of Asia east of the Ganges and were wholly unaware of the New World. Not surprisingly Bacon held that excessive respect for antiquity would necessarily obstruct the advance of science and he was fully aware of the importance of the modern discoveries:

. . . by the distant voyages and travels which have become frequent in our times many things in nature have been laid open and discovered which may let in new life upon philosophy. And surely it would be disgraceful if while regions of the material globe—that is, of the Earth, of the sea and of the stars—have been in our times laid widely open and revealed, the intellectual globe should remain shut up within the narrow limits of old discoveries.

Generally speaking there were, he thought, four major hindrances to the advance of knowledge, science and technics. These he designated the 'Idols of the Tribe', 'of the Den or Cave', 'of the Theatre' and 'of the Market Place'. The Idols of the Tribe are the inherent limitations that man suffers because he is human. He is physically feeble, he has poor eyesight compared with a cat or an eagle, he is deficient in the senses of touch, smell and hearing; necessarily therefore he must be handicapped in his attempts to understand and control nature. The Idols of the Den comprise those limitations that men suffer by virtue of their education and the society to which they belong. The Idols of the Theatre comprise those great systems of thought—Aristoteleanism is an obvious example—that control and circumscribe men's thoughts. Finally the Idols of the Market Place are the limitations and ambiguities imposed by the nature of the words and language men have to use in communicating with each other.

* The word 'technology' first appeared in the English language in 1615.

These four Idols constituted in effect functional concepts for a sociology of science and technics. For an actual scientific procedure Bacon proposed a reformed system of induction but much more important from our point of view was his account of the relationship between science, or systematic knowledge, and technics. With the advancement of knowledge must come new opportunities of invention. New discoveries can lead to inventions that would otherwise be inconceivable or that would be ridiculed as quite impossible.

> If for instance before the invention of the cannon one had described its effects in the following manner: there is a new invention by which walls and the greatest bulwarks can be shaken and overthrown from a considerable distance, then men would have started to contrive different means of increasing the force of projectiles and machines by means of weights and wheels and other modes of battering and throwing. But it is unlikely that any flight of fancy would have hit on the fiery blast, expanding and developing itself so suddenly and violently, because none would have seen an instance at all resembling it in the least; except perhaps in earthquakes or thunder which they would then have immediately rejected as the great operations of nature, not to be imitated by men.

In much the same way, argued Bacon, no one acquainted only with textile threads made from animal furs or vegetable fibres could imagine silk threads produced so copiously by the silk worm; nor could anyone without knowing of the remarkable properties of the lodestone conceive of the mariner's compass. It is therefore very probable that many things yet undiscovered and beyond our present imagination may be brought to light and so provide bases for new and very radical inventions.

On the other hand there are many inventions that unlike cannon, silk and the compass do not appear to depend on the scientific properties of things. Printing, for example, 'at least involves no contrivance that is not clear and almost obvious'.

Bacon in fact distinguished clearly between what we may call science-based and empirical inventions. The former depend on the advance of knowledge and can only be made when, for example, the explosive properties of gunpowder, the magnetic properties of the lodestone, the life cycle of the silk-worm and so on are clearly understood. Of course this represents a rather simple standard of 'science'; it is not so much the sophisticated, highly conceptualised science of today as a stock of general knowledge. Nevertheless the distinction between scientific and empirical invention seems valid enough. In the first category we should place such things as the heat engine, nuclear reactors, electronic devices, the aeroplane and television, and though one may question Bacon's over-simplified assessment of the printing press, there can be little doubt that things like barbed wire, the zip fastener, sewing machines, and revolving doors all belong in the second category. But of the inventions Bacon discussed three seemed to be of particular importance:

> Again we should notice the force, effect and consequences of inventions which are nowhere more conspicuous than in those three which were unknown to the ancients; namely printing, firearms and the compass. For these three have changed the appearance and state of the whole world; first in literature, then in warfare and lastly in navigation; and innumerable changes have been thence derived, so that no empire, sect or star appears to have exercised a greater power and influence on human affairs than these three mechanical discoveries.
>
> It will perhaps be as well to distinguish three species and degrees of ambition. First that of men who are anxious to enlarge their own power in their country, which is a vulgar and degenerate kind; next that of men who strive to enlarge the power and empire of their country over mankind, which is more dignified but not less covetous. But if one were to renew and enlarge the power and empire of mankind in general over the universe such ambition (if it may be so termed) is both more sound and more noble than the other two. Now the empire of man over things is founded on the arts and sciences alone for nature is only to be commanded by obeying her.

The penultimate word 'obeying' should be noted.

Bacon thus provided a workable and reasonably valid classification of inventions and also a generous and truly international ideology for the advancement of technics. Further, in his final work, the posthumous *New Atlantis* (1627), he suggested a specific organisation for the encouragement of technics. In the mythical utopia he described there was an institution called Solomon's House, the basic function of which was to realise the goals Bacon had described in his earlier works.

> For the several employments and offices of our fellows: we have twelve that sail into foreign countries, under the names of other nations, for our own we conceal, who bring us the books and abstracts and patterns of experiments of all other parts. These we call merchants of light.
> We have three that collect the experiments which are in all books. These we call depredators.

And so on to the extent of eighteen more salaried fellows whose duties included carrying out experimental investigations, assessing results and using the scientific knowledge gained to make further practical inventions. There were also to be novices and apprentices. Language apart it all looked very modern in spirit, resembling a large research organisation with overseas sales and technical representatives, research and development departments and policy-making executives. Interestingly enough, though, the biggest single group and the first in order of Bacon's fellows was to engage in what can only be called international industrial espionage; this was a reflection, no doubt, of the technically backward condition of England at that time.

Bacon, it has often been observed, missed the distinctive intellectual trend of his century which was, in the scientific field at any rate, the application of mathematics. This blind spot, if that is what it was, prevented him from appreciating the works of Copernicus and the young Galileo. It might also account for his curious omission of the weight-driven clock from his list of notable inventions; in fact, an appreciation of mechanism does not seem to have been among his many gifts. Rather we must assume that his philosophy of technics and of science was basically *organic* rather than mechanical. Hence he could write that to command nature we must first learn to *obey* her; advice that, significantly enough, is often mistranslated as a recommendation that we must first learn to *understand* nature. This is not very surprising for Bacon's sentiment does not commend itself to our mechanical age which sooner than obeying nature bludgeons her into submission with antibiotics, pesticides, fungicides, hard X-rays and all the armoury of our physical—and therefore basically mechanical—sciences. As it happened the mathematicians and to a somewhat less extent the astronomers were, after the first significant impact, among those thinkers who were least affected by the progress of the discoveries. The explorers might bring back new plants, animals, minerals, metals and accounts of hitherto unknown cultures but they did not bring back any new mathematics. But the mainstream of scientific development at this time lay by way of mathematics. It is therefore difficult for us fairly to appreciate the full extent and nature of Bacon's ideas. For the same reason it seems likely that the cultural significance of the great discoveries has, in terms of the history of science generally, been considerably underrated.

The implications of the great discoveries by the Portuguese and Spanish navigators were later described very succinctly by Christopher Wren in his inaugural lecture at Gresham's College in 1657. Referring to the voyages of Columbus and da Gama he says:

> By these and succeeding Voyages, performed by the Circum-navigators of our Nation, the Earth was concluded to be truly globous, and equally habitable round. This gave occasion to Copernicus to guess why this Body of Earth of so apt a Figure for Motion, might not move among the Coelestial Bodies; it seemed to him in the Consequences probable and apt to solve the Appearances & finding it likewise among the antiquated Opinions he resolved upon this occasion to restore Astronomy. And now the Learned begin to be warm, the Schools ring with this Dispute; all the mathematical men admire the Hypothesis, for saving Nature a great deal of Labour, & the expence of so many Intelligences for every Orb, & Epicycles; yet the apparent Absurdity of a moving Earth makes the Philosophers contemn it, though some of them, taken with the Paradox, begin to observe Nature, and to dare to suppose some old Opinions false and now began the first happy Appearance of Liberty to Philosophy, oppress'd by the Tyranny of the Greek and Roman monarchies.

One practical consequence of the great discoveries and the development of overseas colonisation and trade that followed was the establishment in England of a class of 'mathematical practitioners', or skilled instrument makers who supplied seamen and shipowners with sextants, quadrants, telescopes, compasses, etc. Later still, in the eighteenth century, these men played a very important part in the rise of engineering and the industrial revolution.

GALILEO

Galileo Galilei (1564–1642) was one of the leaders, perhaps the leader, of the 'mechanical philosophy' that swept Western Europe in the course of the seventeenth century. His contributions to technology are usually enumerated as the inventions of the pendulum clock and the astronomical telescope. But in fact they were deeper and of much more general significance than these. His scientific method has often been compared, favourably, with Bacon's reformed inductive procedure, but it is no less instructive to regard Galileo and Bacon as complementary rather than contrasting figures. If Galileo provided the correct, and therefore fruitful, insights at least in so far as the mechanical sciences and technology are concerned, Bacon described the social framework within which these ideas could be made increasingly effective.

The essence of the Galilean method is easily described. It rested on a Platonic belief that the underlying laws of nature are mathematical but that their essential simplicity is hidden from casual observation by the complexities caused by particular and local conditions. Thus the basic simplicity of the principle of inertia and the law of fall of bodies were for long concealed from our understanding by things like friction and air resistance. Imagine these eliminated and the true essential simplicity of the laws will be apparent. (Plato's theory that the mind, or soul, has access to the world of unchanging Ideas, though all things on earth are imperfect, is relevant at this point.) Having intuited the law the Galilean scientist must then devise an experimental procedure using components that approximate to the ideal as closely as possible: weightless strings, friction-free planes and pulleys and so on. The correctness of the supposed law can then be experimentally tested. A few well designed experiments should be sufficient: one does not need a great mass of evidence as Bacon had supposed.

An advanced procedure like this was, one feels, only possible in a fairly mature and stable society in which an extensive amount of knowledge had already been gained and suitably evaluated: astronomical knowledge, mathematical and mechanical knowledge, geophysical, ballistic and military knowledge. In such a society the gaps in knowledge may have become increasingly apparent and more clearly defined. It was the achievement of Galileo to fill these gaps by combining his Platonic faith in mathematics and intuition with the basic techniques of Archimedean mechanics. But Galileo did more than solve the revealed problems; he laid the foundations of the general science of mechanics.

Let us begin by considering the science of machines.

We must suppose that up to, and indeed for some time after, the time of Galileo the operations of machines would be judged according to a few simple commonsense rules. Thus people would assess the quality of a machine by purely normative standards: was it well built and of good materials?, would it serve its purpose with some margin for emergencies?, was it aesthetically satisfying, or even ingeniously and pleasingly complex? The answers to questions such as these would determine the quality of the machine. On a more speculative plane many people regarded machines as ingenious devices for cheating nature, for getting something for nothing. After all, nature herself is very prodigal in the ways in which she dispenses her powers: think of the rivers, the wind and the tides. And not very far from the thoughts of some men would be the old dream of discovering the mechanism of a true self-contained perpetual motion machine.

These and similar beliefs were suitably refuted by Galileo. Probably after centuries of fruitless speculation and inventive effort men, or rather the more sceptical part of men, had come to realise that nature could not be outwitted; there must be some basic principle according to which no one could ever get the better of her. In another field the chemists, encouraged by the Dutch artists and sceptical playwrights like Ben Jonson, had been brought to see that there was no real possibility of converting base metals into gold. Galileo puts the point very clearly:

> The principal cause of disappointment among designers of machines is the belief which these craftsmen have, and continue

to hold, in being able to raise very great weights with a small force, as if with their machines they could cheat nature, whose instinct—nay whose most firm constitution—is that no resistance may be overcome by a force that is not more powerful than it. How false such a belief is, I hope to make most evident with true and rigorous demonstrations that we shall have as we go along.

The correct nature and use of machines is set out:

> . . . perhaps the greatest advantage brought to us by mechanical instruments is with regard to the mover, either some inanimate force like the flow of a river being utilized or an animate force of much less expense than would be necessary to maintain human power—as when we make use of the flow of a river to turn mills, or the strength of a horse to effect that for which the power of several men would not suffice. In this way we can gain great advantage also in the raising of water or making other strong exertions which doubtless could be carried out by men in the absence of other devices. For men can take water in a simple container and raise it and empty it where it is needed; but since a horse or other such mover lacks reason as well as those instruments which are required in order to grasp the container and empty it at the proper time, returning then to refill it, and is endowed only with strength, it is necessary for the mechanic to remedy the natural deficiencies of such a mover by supplying artifices and inventions such that with the mere application of its strength it can carry out the desired effect. And in this there is very great utility, not because those wheels or other machines accomplish the transportation of the same weight with less force or greater speed, or through a larger interval, than could be done without such instruments by an equal but judicious and well organised force, but rather because the fall of a river costs little or nothing, while the maintenance of a horse or similar animal whose power exceeds that of eight or more men is far less expensive than it would be to sustain and maintain so many men.

The function of a machine, then, is to deploy and use the powers that nature makes available in the best possible way for man's own purposes. A machine is an extension of and substitute for human aptitudes and strength. Galileo also sees that the performance, or the job done by a machine is not to be judged by the more obvious signs and symptoms. A machine creaking and groaning under its load is not, in spite of appearances, necessarily doing the most work. By the same token a man who can easily and comfortably cope with his burden may well, indeed almost certainly will, do more work than one who can only just cope. This insight can only have come to a man who had already decided that the criterion is the amount of

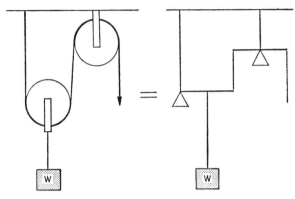

Figure 9.

work done—however that is evaluated—and not a subjective assessment of the effort put into accomplishing it.

There are, essentially, two components of any given machine; the motive agent which can be the forces of water, or wind, or animal muscle, and the mechanism whereby the applied effort is transformed to achieve the desired end. Galileo had little difficulty in showing that the mechanism of all machines can be reduced to a simple system of levers (Figure 9).

Archimedes had enunciated the principle of the lever: for equilibrium the products of the weights and their distances from the fulcrum must be equal (Figure 10):

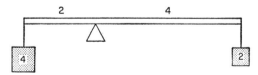

Figure 10.

But it had been impossible to extend this principle to the *dynamic* case when the lever is used as a machine and not as a

static balance. A very simple and obvious generalisation, based on extensive commonsense experience, stood in the way. Let us illustrate it by considering the case of a small crane (Figure 11).

Everyone can confirm from their own past experiences that it is much easier to hold the load, W, in equilibrium than it is to raise it, even very slowly. The conclusion that was drawn from these experiences was straightforward: the 'force' needed for motion must be much bigger than the 'force' needed to maintain equilibrium. The expression is, of course, inexact; it does not specify how much bigger, nor in fact could it, for the inequality varies from case to case. In some instances more 'force' is required to move the load, in others less.

So long as this almost self-evident principle was accepted it was hopeless to look for an exact relationship between the

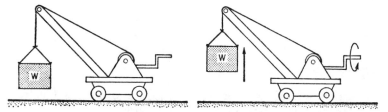

Figure 11.

effort or 'force' put into a machine and the work done by it. Characteristically, Galileo saw through this. The inequality is not really a principle of nature, it merely reflects the imperfections of machines. Basically it arises from the friction of gears, bearings, pulleys etc. together with the other resistances due to motion, all being made worse by the distortions of the machine under load. Writers before Galileo—some of them very able—had seen that friction reduces the performance of a machine but none of them had been willing or able to think the problem right through to its logical conclusion: what must happen with a perfect machine? If we imagine such a machine, totally free from friction, distortion and other forms of resistance then it is easy to see that a trifling addition to the driving 'force' must set the load in (very slow) motion. And once the load is moving it will by the principle of inertia that Galileo had propounded, continue in motion unless and until some resistive 'force' stops it. This can be easily appreciated in the case of the lever, to which all machines can, in principle, be reduced (Figure 12).

A trifling addition of weight, so small that it can in effect be neglected, added to the 'force' or driving side will set the load

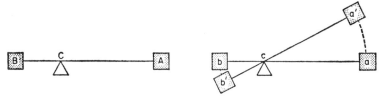

Figure 12.

rising, uniformly and smoothly, if very slowly. Indeed, in the limit we can say that the 'force' required to produce motion is equal to the 'force' required for equilibrium:

> . . . and since to make the weight B descend, any minimal heaviness added to it is sufficient, we shall leave out of account this insensible quantity and not distinguish between the power of one weight to sustain another and its power to move it.

We can now extend the principle of the lever from the static to the dynamic case. Since bb' has the same ratio to aa' as BC has to CA we can write, for the dynamic case, $B.bb' = A.aa'$. But the weights B and A travel over the distances bb' and aa' in the same time so our equality becomes B × velocity of $B = A$ × velocity of A, or $B \times v_b = A \times v_a$. The product of a weight, or load, multiplied by its velocity is, as it happens, the measure of the power being exerted.

This profound insight can hardly be underestimated. It means that with any particular agent the effect, provided we can ignore the frictional losses, must always be quantitatively the same no matter what machine it is applied to. Harnessed to a coal whim a strong horse can hoist a load of 330 lb. of coal 100 feet up a mine shaft in one minute; harnessed to a pump the animal can lift 660 lb. of water some 50 feet to a roof tank in one minute; working a small crane the horse can raise a 3,300-lb. slab of masonry 25 feet in two and a half minutes. The 'force' applied is, in each case, the same and by Galileo's principles, the work done in each case is also the same, being equivalent to 33,000 lb. raised 1 foot in one minute. (This, in fact, is the measure that was later standardised by James Watt as the power of a horse (1784).)

Not only did Galileo's insight make possible the ultimate quantification of power and the establishment of a science of machines but it also opened the way for the elaboration of such concepts as work and energy that were to become of fundamental significance in physics as well as in engineering. Before Galileo the norms were qualitative, after Galileo they became quantitative. Franz Reuleaux made an extremely penetrating observation when he wrote:

> In earlier times men considered every machine as a separate whole, consisting of parts peculiar to it; they missed entirely or saw but seldom the separate groups of parts which we call mechanisms. A mill was a mill, a stamp a stamp and nothing else, and thus we find the older books describing each machine separately from beginning to end. So for example Ramelli (1588) in speaking of various pumps driven by water-wheels describes each afresh from the wheel, or even the water driving it, to the delivery pipe of the pump. The concept 'water-wheel' certainly seems tolerably familiar to him, such wheels were continually to be met with, only the idea 'pump'—and therefore the word for it—seems to be absolutely wanting. Thought upon any subject has made considerable progress when general identity is seen through the special variety;—this is the first point of divergence between popular and scientific modes of thinking.

It is salutary to reflect that the word and concept of 'pump' can have presented such difficulties at the end of the sixteenth century. But it is perhaps less surprising when we recall that apart from one or two points of similarity there was little or no common ground between machines which served different purposes. How could one possibly compare a fulling stocks with a corn mill, a saw mill with a blast furnace, a mine pump with a pump for supplying a mansion with water? All of them did entirely different things and the only common question to ask was whether each machine served its purpose well. The answer, as we remarked, could only be normative: it was a good machine or it was not. But according to Galileo's arguments all machines, no matter what purposes they serve, have the common function of transmitting and applying 'force' or power as efficiently as possible and, moreover, the performance of machines can be quantified, for ideally the product of the driving 'force' and its velocity equals the product of the load multiplied by its velocity. Once all this has been accepted a rational science of machines becomes possible and the design and function of each component can be studied without regard to the final purpose which the machine serves. That is, without considering whether it grinds corn or saws wood for it is the quantity of work done

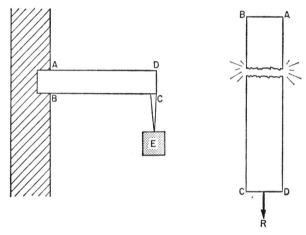

Figure 13.

that is important and to which both can, in the last resort, be reduced.

The working out of Galileo's ideas in their bearing on technology was begun in the seventeenth century and continued in the eighteenth century. Various measures of the 'effect', or work capacity, of machines like water-wheels or steam engines were used or proposed by engineer-scientists like Mariotte, Parent, Desaguliers, Beighton and finally culminated in Watt's standardised 'horse-power' of 1784. The establishment of a science of machines was rather more gradual and, according to Reuleaux, really got under way only with the work of the men associated with the Ecole Polytechnique (founded in 1794). But laying the foundations for these developments was not the only contribution made by Galileo to the establishment of scientific technology in place of technics, crafts and empirical invention. He initiated the science of the strength of materials. Once again the fruitful principle of the lever is applied to a technological problem: the determination of a general expression for the strength of *any* beam carrying a load (Figure 13).

> Let us imagine a solid prism ABCD fastened into a wall at the end AB and supporting a weight E at the other end . . . the wall is vertical and the prism or cylinder is fastened at right angles to the wall. It is clear that if the cylinder breaks fracture will occur at the point B where the edge of the mortise acts as a fulcrum for the lever BC to which the force is applied; the thickness of the solid BA is the other arm of the lever along which is located the resistance. This resistance opposes the separation of the part BD lying outside the wall from the portion lying inside. From this it follows that the magnitude of the force applied at C bears to the magnitude of the resistance found in the thickness of the prism, i.e. in the attachment of the base BA to its contiguous parts, the same ratio which the length CB bears to half the length BA; now if we define absolute resistance to fracture as that offered to a longitudinal pull . . . then it follows that the absolute resistance of the prism BD is to the breaking load placed at the end of the lever BC in the same ratio as the length BC is to the half of AB in the case of the prism or the semidiameter in the case of a cylinder . . . the weight of the solid BD itself has been left out of consideration, or rather the prism has been assumed to be devoid of weight. But if the weight of the prism is to be taken into account in conjunction with the weight E we must add to the latter one half that of the prism BD so that if, for example, the prism weighs two pounds and the weight ten pounds we must treat E as if it were eleven pounds.

We can perhaps illustrate rather more clearly what Galileo has done by setting out the lever system he refers to as the actual bare bones of the problem. By abstraction he has reduced the problem to its basic elements (Figure 14). If we know the longitudinal strength of a beam of any material we can easily calculate what load it can bear when that load is applied transversely and uniformly distributed. The solution is entirely general and it is plainly of great importance for architects and builders concerned, for example, with calculating floor loadings. From it one can easily deduce that since, at the breaking load E, we have

$$R. \tfrac{1}{2}AB = E. BC, \quad \text{or} \quad E = R\frac{AB}{2BC}$$

it follows that to support as big an E as possible AB must be as big as possible. This is commonly achieved by the familiar 'I' section of a girder.

There is, as it happens, an interesting error in Galileo's

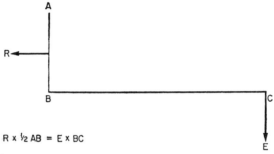

R × ½ AB = E × BC

Figure 14.

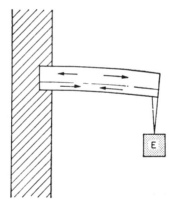

Figure 15.

reasoning: he did not allow for the elasticity of the beam. In fact the beam is not all stretched; the lower layers are actually compressed and bending takes place about a fulcrum that is inside the beam (Figure 15). This error on Galileo's part was subsequently corrected by later writers who were, directly or indirectly, disciples of his. But the mistake that he made does not deprive Galileo of the credit of first pointing out the correct method of tackling this problem.

Thus Galileo made revolutionary contributions not only to the theories of machines and of power but also to the new and basic technology of the strength of materials and the theory of structures. In paying just tribute to this quite remarkable man we must never forget that he was, in fact, building on the triumphs of the great medieval and renaissance inventors and mechanics.

THE SCIENTIFIC REVOLUTION AND THE PROGRESS OF TECHNOLOGY

The 'century of genius', as A. N. Whitehead aptly called the seventeenth century, witnessed not only the 'scientific revolution' but a number of other dramatic changes as well. Germany and Italy, who had contributed so much to science and technology, lost their pre-eminence and by the end of the century the torch of leadership had passed to France, the Low Countries and England. Scientific activity in Germany remained at a low level throughout the eighteenth century and did not begin to revive until the second and third decades of the nineteenth century. Able Germans, born in this period of national decline, usually sought recognition and fortune abroad, as for example (Sir) William Herschel did when he came to England.

The causes of the decline of German and Italian science were social, political and economic and therefore outside the scope of our discussion. A change for the better, and one which was not unrelated to the scientific movement, was the abrupt decline in belief in witchcraft and the ending of the persecutions that were associated with it: the dark side, as it were, of the occult sciences of the middle ages. At the beginning of the seventeenth century witch trials took place in both France and England and in the latter country there was a particularly bad outbreak of persecution in the middle of the century. But then the climate suddenly changed and belief in witchcraft vanished like the evil dream it was. The England of the Royal Society (1660–1662), of Restoration comedy, of King Charles the Second was not a place in which such fantasies could flourish and though it lingered on into the eighteenth century in the more remote places like Scotland and America its days were numbered in those countries too.

The scientific movement was rather more complex and rather less coherent than the expression 'scientific revolution' might suggest. It owed a good deal to Galileo and his work can be taken as representative of many of the main strands. Thus the new philosophy of nature, expounded in the most comprehensive and systematic way by René Descartes (1596–1650), is foreshadowed in the writings of Galileo. According to this philosophy all natural phenomena are merely the effects produced by inert bodies moving according to immutable mathematical laws. The sensation we call heat, for example, was thought to be due to the vibrations, or to the characteristic shapes, or perhaps to both, of certain specific atoms; the sensation of light was held to be due to the impact of other sorts of atoms on the retina of the eye, or else to vibrations transmitted through an all-pervading aether. According to this philosophy all things in nature that were not mathematical or geometrical, like the shapes, sizes, motions of the atoms of which all bodies were comprised, were to be regarded as mere illusions of the senses. There was no colour in nature—no blue in the mid-day sky, no red at sunset—no scents or sounds, no warmth; merely matter in endless motion.

Whatever the philosophical difficulties of the new metaphysic—and generations of philosophers have earned their bread and butter by examining these difficulties—there can be no doubt that it provided a convenient and ultimately very fruitful set of axioms for the development of the physical sciences. Indeed, it was within the framework of the 'mechanical philosophy' that modern physics and chemistry developed so triumphantly and it is probably true to say that most working scientists accepted (and still accept) its main tenets. It is, however, less well suited to the biological, psychological and social sciences.

The choicest fruits of the mechanical philosophy were not to be gathered until the nineteenth century when John Dalton succeeded in bringing the previously speculative concept of the atom within the realm of positive science and when the undulatory theory of light, the electromagnetic field theory and the comprehensive doctrine of energy came to be established. During the seventeenth century the most successful application of the mechanical philosophy lay in the development of the system of mechanics pioneered by Galileo and the final solution, by Isaac Newton, of the immemorial problem of the motions of the heavens (*Principia mathematica et philosophia naturalis*, 1687).

Kepler had shown that the planes of planetary orbits all pass through the sun and he had established that the planets move in ellipses, the sun being at one focus, that the line joining the planet to the sun covers equal areas in equal times and, lastly, that the square of the periodic time (the planetary year) is proportional to the cube of the mean radius of the planet's orbit. At roughly the same time Galileo had shown by means of an astronomical telescope that the planets are all like the earth: globes made up, in all probability, of 'earthy' material. The problem of the heavens was posed with unprecedented clarity; it was seen to be an experiment, as it were, set out by some celestial laboratory attendant. Granted that inert bodies like projectiles and planets follow strictly determined paths when not subjected to such local complications as friction and air resistance, what basic principles can we find to account for their motion? Newton solved the problem. He did so by accepting the principle of inertia in the form Descartes had given it; by postulating the ideas of mass, accelerative force and momentum; by proposing the third law of motion (action and reaction are always equal and opposite); and lastly by putting forward the exact concept of universal gravitational attraction. These immensely powerful intellectual tools were enough for him to reduce the whole complex problem of the 'system of the world' (the solar system) to the same basic principles as those that determine the motions of things like projectiles and pendulums on earth. The distinction between heaven and earth had been qualitatively destroyed by Galileo's observations with the astronomical telescope; it was quantitatively, or formally, destroyed by Newton's grand synthesis.

And yet Newton's *Principia* was in a sense a conservative work; it marked the end rather than the beginning of an epoch. Thereafter planetary astronomy ceased to be a matter of urgent inquiry and became a specialism; a matter for a small and select group of skilled professional civil servants concerned with nautical almanacs, calendars and standard times. The attention of speculative astronomers turned to sidereal astronomy, for the first great riddle of the universe had been solved.

Newton's prestige was, and is, unassailable but on one point of his system at least disagreement was possible from the beginning. Newton had taken as the 'quantity of motion' of a body the product of the mass and the velocity, or the momentum. This was entirely reasonable for a body's capacity to give motion to another body by collision is proportional to its velocity. One billiard ball striking another will cause the second one to move with a speed which is proportional to the velocity of the first. Moreover, if we take the velocity of a body moving in one direction to be a positive quantity while that of a body moving in the opposite direction is a negative quantity it can easily be shown that the total quantity of motion is always conserved in collisions; that is, it is the same before and after impact. On the other hand Huygens and Leibnitz took the quantity of motion to be as the *vis-viva*, or the product of the mass multiplied by the *square* of the velocity. This, too, was reasonable for the capacity of a moving body to overcome resistance is proportional to the square of its velocity. That is, its ability to penetrate soft clay, earth, sand, wood or any other substance, or to rise upwards against the force of gravity increases with the square of its velocity. But as it happens *vis-viva* is conserved in perfectly elastic collisions only; in all other cases there is a loss of *vis-viva* when bodies collide.*

A dispute between those who favoured the Newtonian measure of quantity of motion and the followers of Huygens and Leibnitz went on in desultory fashion throughout the eighteenth century. It is sometimes said that it was all a dispute about words since the same results could be deduced from either theory and there was therefore no real inconsistency

---

* If two identical billiard balls travelling with the same velocity but in opposite directions collide, the total momentum, or quantity of motion, before the collision is $m\,v + m\,(-v)$ which is equal to 0. After the collision and assuming imperfect elasticity, it is $m\,(-v') + m\,v'$ which again equals 0. Momentum is thus conserved. If we have two identical snowballs in collision momentum is again conserved but *vis-viva* is not, for $m\,v^2 + m\,v^2$ equals $2m\,v^2$, before collision, while after collision the *vis-viva* is 0 as the two balls coalesce and fall to the ground without velocity.

between them. But in the long run it was more than a semantic quarrel for the concept of *vis-viva* led to deeper and more general insights than its rival; in fact it developed, side by side with the engineer's measures of 'effect' or work, 'duty' and 'power' into the generalised and powerful concept of energy. As it happened the notion of *vis-viva* was from the beginning readily applicable to machines; for, after all, machines are concerned with overcoming resistances. Newton, we may add, was unlike all his great contemporaries in that he was professedly uninterested in technology.

THE EXEMPLARY EXPERIMENT

The last important consequence of the great scientific movement that we shall mention was the rise of systematic experimental science. This should not surprise us for systematic experiment is hardly possible without some guiding structure of axioms and principles of the sort Galileo, Descartes and their contemporaries erected. One of the most instructive experiments of the period, indeed in the whole historical record of physical sciences, was carried out by Newton when he established the compound nature of light.

It had been shown that when light passes through a transparent substance like glass or water it is deflected from its original course to an extent that can be found by means of a simple rule called Snel's law. Now Newton noticed that when light is passed through a prism so that it produces the familiar rainbow-like spectrum this law appears to break down for some of the light is more deflected than the rest; sunlight coming from a round hole in the blind produces an elongated image in the spectrum colours. This break-down in scientific law intrigued Newton and he resolved to find the cause. He tried substituting different prisms to see if the quality of the glass had anything to do with it; he varied the size of the hole in the blind and he tried sending the light through different thicknesses of the prism. In short he examined one by one all the possible causes of the anomaly. In each case the result was negative. Finally he realised that the cause must lie in the nature of light itself. Guided by his scientific intuition he isolated the red part of the spectrum and caused the light ray that produced it to pass through a second prism. This time the image was not another spectrum but simply and solely red in colour. Nothing he could do to the light that produced red illumination could extract another colour from it; it remained obstinately monochromatic. The same was found to be true of the other colours and shades of the spectrum. Newton therefore concluded that 'white' light is made up of different rays that are differently deflected, or more precisely, refracted by a prism and these rays are elementary in that they cannot be split up into simpler components. The action of the prism is, therefore, to analyse white light into its constituent rays according to their different refrangibilities. This experiment was not, as we shall see, without its technological consequences.

THE DISCOVERY OF THE ATMOSPHERE

The mechanical philosophy, judged by its more esoteric achievements could only be appreciated by men of the very highest intellectual calibre. But one of its off-shoots that was much more readily comprehensible, in many ways more immediately dramatic and hardly inferior in the importance of its consequences was the discovery of the true nature of the atmosphere. The acceptance of the Copernican system coupled with the manifest absence of either retardation of the motions of the earth and other planets or of an irresistible wind blowing from the east indicated clearly enough that the 'air' must be finite and local to the earth. Between the sun, the planets and the satellites there must be empty space, devoid of air. Galileo had already treated air as an Archimedean fluid in his studies of the law of fall of bodies; his disciples Torricelli and Viviani took his reasoning to its logical end and concluded that the air must comprise a finite ocean which must therefore exert a pressure on all objects at the bottom of it, just as the more familiar oceans and seas do. They reasoned that it was this 'atmospheric' pressure that explained the operation of an ordinary suction pump and the fact that such pumps could not raise water a distance greater than 32 feet was simply because the weight of 32 feet of water equalled the pressure of the atmosphere. From this they went on to invent the mercury barometer prompted by the reflection that as mercury is thirteen times heavier than water the pressure of the atmosphere could support a column of mercury only $32 \times \frac{12}{13}$ inches, or 30 inches high. Later on Blaise Pascal initiated a famous series of experiments (1648) in which Périer showed, by taking a barometer up the Puy-de-Dôme, that the pressure fell appreciably. By correlating the fall of pressure in inches of mercury with the known height of the mountain Périer was able to compute the height of the atmosphere; or, perhaps more exactly, the depth of the ocean of air round the earth.

The discovery of the fact that we all live at the bottom of an

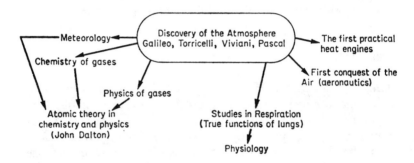

Figure 16.   *Consequences of the discovery of the atmosphere.*

ocean of air that exerts a pressure of some 15 lb. per square inch on us, and everything else on the surface of the earth must rank as the most bizarre discovery ever made. Certainly it intrigued contemporaries much more than the rather remote triumphs of Newtonian mathematics and mechanics. Nor was this interest unjustified for the discovery proved immensely fruitful. Its consequences are represented in diagrammatic form (Figure 16).

With the bulk of these we are not directly concerned, but the two on the right are of very direct interest. We may put the point briefly: for countless centuries men had entertained two ambitions, firstly to master the air and fly in the way in which he had mastered the seas with his ships; and secondly, to master the evident powers of fire. Now at last, if indirectly, the sovereign discovery of the ocean of air surrounding our globe, perhaps the last and possibly the greatest of the 'discoveries', was to enable men to achieve these two ancient ambitions.

Suggestions for flying machines had, from ancient legends down to the more sophisticated ideas of the renaissance Italian artists and engineers, all been based essentially on the principle of imitating the birds—what other method, it might have been asked, was conceivable?—and all had failed simply because it was impossible to get enough power to lift the weight of a man off the ground for any appreciable time. The problem of man-powered flight has still not been solved so it is apparent that early would-be aviators did not contribute significantly to the final conquest of the air.

If, however, we recognise that the air is an Archimedean fluid physically different from water only in that it is lighter and happens to be compressible, then we can conceive of a boat to float on or in the air just as ordinary boats float on or in water. The principle of displacement, established by Archimedes, being common to all fluids, it follows that an 'air boat' must be one whose bulk weighs less than the same volume of air.

The first known suggestions for such a boat were put forward in 1670 by the Jesuit, Father Francesco Lana Terzi. Father Lana's boat consisted of four large copper spheres, very thin and evacuated of air. These vessels would therefore weigh less than the surrounding air and should have enough positive buoyancy to lift an attached gondola with crew. Of course the proposal was quite impracticable: the copper spheres could not be made at once thin enough to have negligible weight and strong enough to withstand the pressure of the air. But the basic idea proved sounder than the previous schemes for 'heavier than air' flight, for it was in line with Father Lana's proposal that the first successful balloons using very light gases such as hot air, or later hydrogen, to give positive buoyancy were made in the eighteenth century. This was a development that owed a good deal to the rise of 'pneumatic' chemistry and the physics of gases.*

Could the simple hot air or hydrogen balloon have been invented empirically? The answer is almost certainly no. It is

---

* Among the pioneers of ballooning in the later eighteenth century were such well-known chemists and physicists as Montgolfier, Gay-Lussac and J. A. C. Charles.

one thing to know that hot air rises—as indeed Aristotle remarked—but it is something quite different to interpret this
phenomenon in Archimedean terms and to calculate that it
should be possible to harness the positive buoyancy of a large
volume of hot air, or of hydrogen, to lift not merely the
fabric of a balloon but also the gondola, its shrouds and the
'pay-load'.

The other major technological consequence of the discovery
of the atmosphere, the invention of the first practicable heat
engine, was more immediate and more important in its impact;
it led directly to the evolution of all the other kinds of heat
engines, including the particular type that was to make the
heavier-than-air flying machine—the aeroplane—a feasible
proposition at the beginning of the twentieth century. The
distinctive line of development began in this case with the work
of Otto von Guericke, the man who was, perhaps, the last of
the great German scientist/technologists before the long intellectual recession that lasted until the early nineteenth
century.

Von Guericke had succeeded, some time in the 1650s, in
making an air pump; with it he had been able to exhaust
quite large vessels and to study the properties of vacua. We
need not concern ourselves with this aspect of von Guericke's
work, save to remark that he set a fashion in experimentation
and gave a distinctive direction to research that was to last for
over a hundred years.* More significant, from our point of
view, were the chapters in the remarkable book, *Experimenta
nova Magdeburgica de vacuo spatio* (Amsterdam 1672), in which he
described—with excellent illustrations—some particularly intriguing researches on the pressure of the atmosphere.

He put two small hemispheres together to form a sphere and
by means of the air pump, or 'syringe' as he called it, evacuated
the air. He showed, and the illustration was particularly
graphic and convincing, that sixteen strong horses were unable
to overcome the pressure of the atmosphere and pull the
hemispheres apart. He showed too that if a piston (as we
should call it) was put into a cylinder and the air underneath
the piston pumped out then teams of strong men were unable

---

* See Joseph Wright's painting 'An Experiment with the Air-Pump'
(1768). It hangs in the Tate Gallery.

to pull the piston out again against the pressure of the atmosphere.

Perhaps the most intriguing and suggestive of his experiments was the one in which he showed how a boy of only twelve to fifteen years of age could, by means of the air pump, so utilise the pressure of the atmosphere that he could cause a very large weight to be lifted (Figure 17). Once again the 'syringe' was used to evacuate the air from a large brazen cylindrical vessel (*ahenum*), fitted with a round plug (*embolum*)

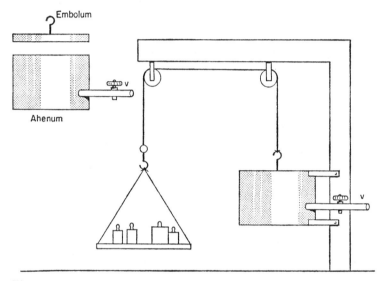

Figure 17.

or, as we should say, piston. The piston was connected via a rope passing over two pullies to a strong wooden platform on which weights amounting to 2,686 Magdeburg pounds were placed. When the air was pumped out through the valve (*epistomium*), *v*, the piston moved down smoothly and lifted the platform with its load.

Von Guericke had therefore demonstrated convincingly the immense 'force' of the atmosphere and had hinted very clearly at a method whereby it* could be harnessed. In the same year,

---

* The different ideas underlying the words force, power, pressure were still not distinguished at this time.

1672, we find the Dutchman Christisan Huygens and the French engineer Hautefeuille proposing, independently of one another, a method whereby this could be done. Instead of using an air pump the explosion of a small charge of gunpowder might, they suggested, produce a vacuum by expelling the air through an outwards opening valve in the piston. Unfortunately it proved impossible to get a good vacuum this way and there were other serious drawbacks to the method, notably the difficulty of igniting the gunpowder. But in 1691 another Frenchman, Denis Papin came up with a suggestion that in principle solved the problem. Instead of gunpowder the cylinder had a pool of water at the bottom. The water was boiled by putting a fire under the cylinder and the piston rose on the steam that was generated. When the piston had reached the top of the cylinder the fire was removed; the cylinder cooled down and as the steam condensed the pressure of the atmosphere drove the piston down.

Thus by 1691 the problem had been solved in theory and it remained only to solve it in practice. At long last by combining the expansive properties of steam with the recently discovered pressure of the atmosphere a way had been found of harnessing the power of fire. From the early experiments of Hero of Alexandria through to the schemes of seventeenth-century inventors like Branca, de Caus, the Marquis of Worcester and David Ramsay, men had sought a way to harness the power of fire but without knowledge of atmospheric pressure it had eluded them. The solution, admittedly still theoretical, had sprung in the first place from the mechanical achievements of medieval and renaissance Europe and from the introduction and development of specific items such as pistons, cylinders and valves. In the second place it owed a good deal to the experimental philosophy of Italians like Galileo and his disciples, Frenchmen like Pascal, Hautefeuille and Papin, the Dutchman Huygens and the German Otto von Guericke. The English who were the first to apply the new-found power were not, at this stage, active in the field of technological innovation.

A common and dearly held belief among the English is that they are and have been the true and original innovators, but that unscrupulous and much less original foreigners often steal English ideas which they then proceed to exploit for their own profit. Curiously enough very similar beliefs are held by Frenchmen, Germans, Russians and no doubt many others* about their original ideas and the ways in which greedy foreigners (including perhaps the English) steal them and turn them to profitable account. However regarded, this patriotic belief is not particularly creditable to those who entertain it. It is therefore refreshing to note that at the beginning of the eighteenth century Daniel Defoe was willing cheerfully to confess that the English were unoriginal and were best at exploiting other people's ideas:

> It is a kind of Proverb attending the Character of English Men, that they are *better to improve than to invent*, better to advance upon the Designs and Plans which other People have laid down than to form Schemes and Designs of their Own; and which is still more, the Thing seems to be really true in Fact, and the Observation very just.
>
> Whether this Reproach upon them is raised upon the Suggestions of Foreign Observers, or whether it be our own upon ourselves, is not worth while to examine; it seems that we are very willing to grant the Fact.
>
> Even our Woollen Manufacture itself, with all the admirable Improvements made upon it by the *English* since it came into their hands is but a building upon other Foundations, and improving upon the Inventions of the *Flemings*; the Wool indeed was *English*, but the Wit was all *Flemish*. . . .
>
> We had the Wool, but understood neither how to comb it or card it, spin it or weave it. . . .
>
> Thus we were all said to learn the Art of building Ships from the *Genoese* and *French*. . . .

Before we conclude that the English, who were admittedly the first to develop and use the steam engine, had merely appropriated an invention whose basic principles had already

* 'The Frenchman can devote himself passionately to creative work, to invention, and then, often enough, he shows no interest in its subsequent application. He sows and others reap. That explains why France is often found at the beginning of things, the motor car and the aeroplane for example, but not always when the profits are shared'; thus André Siegfried, *The Character of Peoples* (London, Jonathan Cape, 1952), p. 55. Substitute Englishman and England for Frenchman and France and these words could have been uttered by any English politician or public figure of recent years.

been worked out on the Continent we would do well to recall two points that we noted above. The first is that a society that is passing through, or has just passed through an imitative phase may well be on the point of becoming a technologically original and creative society. The second is that, as the example of Johan Gutenberg and his printing press demonstrated, the solution of a problem in practice may be no less difficult, no less original that is, than the solution in principle. And this, in the case of the steam engine, was to prove abundantly true.

The first 'fire-engine' to pass beyond the stage of a suggestion on paper and to transcend the scale of a working toy was described and built at the end of the seventeenth century by Captain Thomas Savery, FRS. The engine was patented in 1699 and Savery gave an account of it in a short pamphlet entitled *The Miner's Friend*, published in 1701. Basically his engine was very simple, being nothing more than a double-acting pump in which the lifting or suction effect of the piston in the pump barrel was provided by condensing steam; while the pushing or forcing effect was provided by the application of steam under pressure. A pump barrel, made of metal, is filled with steam and the supply then cut off; cold water is poured on to the barrel and condenses the steam, a vacuum is formed and water rises up through a pipe to fill the void. When the barrel is full of water the steam pressure is turned on again and the water is driven out, through a rising pipe, to a tank higher up. Two upward-opening valves prevent the water falling or being forced down the pipe to the well or reservoir at the bottom. The action was push–pull and was clearly derived from a simple and fairly common pump arrangement converted to steam operation. Savery was, of course, quite aware of the properties of the vacuum and the pressure exerted by the atmosphere.

Just how efficient this device was must be entirely problematical. The waste of heat due to the hot steam condensing on cold metal and on the surface of the cold water must have been very great and it is doubtful if the engine really worked satisfactorily. There were other grave drawbacks, too. Steam under pressure was difficult and dangerous to control in those days and the overall lift that the engine could manage was therefore very limited. Boilers, pipes, valves and 'pump barrels' could not stand steam pressures of more than a very few atmospheres and as a pressure of one atmosphere meant a lift, for water, of only 32 feet the most that an engine could be expected to achieve was about 100 feet or so. But Savery was, we infer, a 'projector' living in an age of projectors; the age that culminated in the South Sea Bubble. He was not, by later standards (including our own), a modest man:

> Miner: Sir, having been some time concerned in the engines now used for drawing water out of our mines and so much talk of this wondrous invention of yours by raising water by fire, I was very desirous to enter into some discourse with you concerning the nature, use and application of your engine, so strangely differing from all other engines ever yet invented for our works and which, you positively affirm, will every way tend so much to our advantage in the use of them and I do not doubt of meeting with that plainness, freedom and good humour, that your discourse is generally accompanied with. And with the same freedom resolve me in such questions as the general sense of us miners may naturally propose to object against the use of your engine, especially such of us as are yet ignorant of its use and operation, who are more capable to judge of fact, than of the nature and power of that which raises your water.

A man who could readily imagine others addressing him in this way was obviously not suffering from a surfeit of modesty about his invention. When one considers the shortcomings of this engine and reflects that there is no evidence that it ever worked reliably and economically in competition with established modes of generating power, then it must seem surprising that Savery has been treated with such great respect by historians and his invention taken at the inflated value he put on it. But whatever the assessment of Savery's engine there can be no doubt that he gave enormous publicity to the possibility of the new form of heat engine and in doing so almost certainly helped the acceptance of the ultimately successful Newcomen engine of 1712. The acceptance of revolutionary innovations in technology, and in science too for that matter, is not always a story of virtue bringing its own reward; a good deal of publicity can make all the difference.

Chapter 3

The Eighteenth Century

In seventeenth-century England a growing demand for fuel had led, inevitably, to an increasing use of coal not only for domestic heating and cooking but also for various chemical and metallurgical processes. In fact there was something of a fuel shortage and, among other things, this stimulated ironmasters to seek ways of using coke instead of charcoal for smelting iron. It was a difficult business but in 1709 Abraham Darby of Coalbrookdale succeeded in using local coke to smelt iron and thus laid the foundation stone of the great iron and steel industries of the present.

In the eighteenth century the shortage that was to prove most significant, technologically and industrially, was that of power. Put simply the problem was how to get the most out of the limited natural power resources available in England, a country much less well endowed with rivers than France or Germany. And the power shortage was a problem that was to grow steadily more acute in certain industries as the century progressed. In response to the shortage Galilean mechanics was to assume an increasing importance and ultimately to lead to the establishment of power technology on an entirely new basis. Let us, at this point, recapitulate the three main contributions made by Galileo:

(1) He demonstrated that all machines can be reduced to the principle of the lever.

(2) He realised that when all imperfections have been eliminated the 'effort' required to produce sustained motion in a machine is equal to that required to keep the load in equilibrium.

(3) He established the principle of inertia: when all resistances are eliminated a moving body, or bodies, will remain in uniform motion indefinitely.

In traditional technics the utility of a stream or river would have been assessed qualitatively: had it a good, steady flow and was it reasonably constant all the year round? In post-Galilean technics the yardstick was *quantitative*: how much natural 'effort' did it have and what proportion of this could we hope to convert into useful 'effect' by means of a suitable water-wheel? An absolutely perfect water-wheel, laid up in some Platonic-Galilean heaven, could conceivably give an effect equal to the effort put in.* But certainly in practice, and no doubt in theory too, there would be good reasons why it would be impossible to transform all the natural effort into useful effect.

From the second half of the seventeenth century onwards, engineers were generally agreed that the effort and the effect could, in all cases, be measured by the raising of a weight through a given distance in a given time. This reflected not only the acceptance of Galilean principles but also the importance of mining technology as the pace-setter at this time. Apart from excavating and transporting the mineral the main problem of mining was to pump hundreds of tons of flood water from the bottom to the top of the shaft every twenty-four hours. The engineers' measure of effort was, therefore a reasonable one to choose.

The natural effort of a stream was easily computed; it was nothing more than the capacity of the stream to lift itself to a given height if deflected upwards to make a fountain. The height of the fountain depended on the velocity of the stream† and the weight lifted is equal to the quantity of water flowing in the stream.

Although we can reduce all machines to systems of levers, the pressure or 'weight' exerted on the moving part of an engine driven by water—or wind, or muscle for that matter—is unlike gravitational weight for the latter does not diminish the faster the machine moves. When an express lift descends with uniform speed a load on its floor, or a suitcase in a passenger's

* That is, in modern terms, that the work done by the waterwheel is equal to the energy of the stream, or of that part of the stream that acts on the wheel.

† As Torricelli showed, the height to which the water will rise is proportional to the square of the velocity of the stream; in modern usage, $v^2 = 2gh$.

hand, exerts the same weight as if the lift was stationary. There seems to be no limit to the speed that gravity can build up; or to the effort that it can exert. On the other hand a man cannot exert any pressure, or 'weight', on a barrow that moves as fast as he can run.

Somewhere between the load that is so great that it stops the machine and one that is so small that the load moves as fast as the (unimpeded) driving agent can, there must be a load and a speed of ascent that, taken together, comprise the maximum useful effect.

Since the motion of the machine necessarily reduces the pressure, or 'weight' of the driving agent and since, in the cases

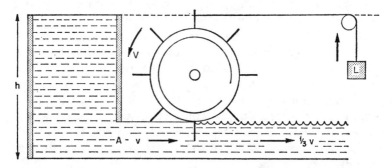

Figure 18.

of wind and water, the latter must leave the machine with some residual speed it seems to follow that no water-wheel, for example, can convert all the natural effort of a stream into useful effect, even if we neglect friction and air resistance. The problem, then, is to calculate the fraction of natural effort of the stream that a perfect water-wheel can convert into useful effect.

These were the considerations that led Antoine Parent to formulate, in 1704, his theory of the greatest possible perfection of machines. Although the theory was intended to be general, its first and most important application was to water-wheels. The pressure or 'weight' that a stream of water exerts against the blade of a water-wheel (Figure 18) is proportional to the relative velocity of the stream and to the quantity of water actually striking the blade. But the quantity of water, too, is proportional to the relative velocity of the stream and so the 'weight' of water on the blade must be proportional to the square of the relative velocity.*

On Galilean principles the pressure, or 'weight' of water on the blade is equal to the gravitational weight of the load, L, when it is stationary or when it is rising with uniform velocity. We can therefore equate the product of the square of the relative velocity of the stream and the speed of the blade, which in Figure 17 is the speed with which L rises, to the total useful effect derived from the stream. Using the newly invented calculus Parent was able to show that in order to obtain the maximum useful effect the load, L, had to be such that the wheel moved with 1/3 the speed of the stream; and, furthermore, that the maximum useful effect was only 4/27 the natural effort of the stream.

This fraction is surprisingly small, the more so when we reflect that the wheel is supposed to be mechanically perfect. However, Parent pointed out, not unreasonably, that the impact of the water is reduced by the motion of the wheel and that as the stream leaves the wheel with 1/3 of its original velocity it must take an appreciable amount of natural effort with it.

More to the point, there were three limitations that Parent unconsciously imposed on his calculations. Following the scientific custom of the time he thought and wrote in the language of geometrical proportions rather than of physical equalities; he therefore omitted the number 2 in his expression for the height to which a stream of water can rise when deflected upwards to make a fountain. This meant that his value for the natural effort of a stream was twice its true value,† so that his fraction, expressing the efficiency of a perfect waterwheel, was effectively halved. The correct figure should have been 8/27, a much more realistic fraction. The second limitation was rather more subtle and no less revealing. In effect he

---

* That is $(v - V)^2$.

† From the knowledge that v^2 is proportional to h, the height to which the water rises, Parent went on to assume that $v^2 = gh$ and not, as he should have done, $= 2gh$. Hence he concluded that the natural effort of the stream was $v . v^2$, or v^3. It should have been $v . \frac{1}{2} . v^2$, or $\frac{1}{2} . v^3$ (which is equivalent to $\frac{1}{2} m . v^2$).

had considered a water-wheel in which only one blade at a time is immersed in the stream. He therefore concluded that the quantity of water striking the blade must be proportional to the *relative* speed of the stream (it would be zero if the wheel turned at the same speed as the stream). But this was in effect an error of abstraction. As we are dealing with a perfect machine we can suppose that it has as many blades as we like so that every particle or drop of water must strike some blade. This would make the quantity of water acting on the wheel simply proportional to the velocity of the stream and with this amendment the fraction is further changed from 8/27 to 1/2. This proves how difficult the art of abstraction really is and it provides additional, if indirect, confirmation of Reuleaux's observation about the development of the science of machines.

The final limitation was simply that Parent did not bother to consider the performance of an overshot water-wheel; one, that is, in which the water passes over the wheel, filling buckets and so acting by gravitational weight and not by impact against blades. It is probable that he did not consider this case because, as he pictured the process, the result must have been the same in both cases. If, in Figure 18, we imagine the immersed blade detached from the wheel and moved upstream until it blocks the orifice A, thus stopping the flow, the pressure on the blade will be transformed, practically imperceptibly, from that due to impact to that due to the hydrostatic pressure, or gravitational weight, of the water in the reservoir. Why should the gravitational weight of water produce a different result from impact 'weight'? After all both cases are consistent with the relationship that v^2 is proportional to h.

In this Parent happened to be mistaken. There was a fundamental difference in the two cases that was not apparent on his system of analysis.* But it is fair to say that Parent's magnificent achievement—which has never been adequately appreciated by historians—initiated a debate in which the clarification of

---

* Parent's calculations were transformed into Newtonian mechanics by Colin Maclaurin, the eminent mathematician. The transformation involved considering the *force*, or rate of change of momentum, of the water against the blade instead of the weight or 'pressure'. But both Parent and Maclaurin ignored the *contraction* of the water as it issues from the orifice: the effect known as *vena contracta*.

his points and the correction of his limitations led to the establishment of the true theory of power and contributed significantly to the ultimate establishment of the energy doctrine.

An interesting sidelight on the problem of the efficiency of engines in general and water-wheels in particular was provided by Dean Swift:

> His Lordship added that he would not by any further particulars anticipate the pleasure I should certainly take in viewing the grand academy, whither he was resolved I should go. He only desired me to observe a ruined building upon the side of a mountain about three miles distant, of which he gave me this account: that he had a very convenient mill within half a mile of his house, turned by a current from a large river, and sufficient for his own family, as well as a great number of his tenants. That, about seven years ago, a club of those projectors came to him with proposals to destroy this mill and build another on the side of that mountain, on the long ridge whereof a long canal must be cut for a repository of water, to be conveyed up by pipes and engines to supply the mill because the wind and air upon a height agitated the water, and thereby made it fitter for motion: and because the water descending down a declivity would turn the mill with half the current of a river, whose course is more upon a level. He said that being then not very well with the court and pressed by many of his friends he complied with the proposal; and employing an hundred men for two years, the work miscarried, the projectors went off, laying the blame entirely upon him, railing at him ever since and putting others upon the same experiment, with equal assurance of success, as well as equal disappointments.

It is remarkable that the 'projectors', or rather Dean Swift,* happened to be quite correct in claiming that water 'descending down a declivity' is twice as effective as water 'flowing on the level'. In this respect they were well ahead of Parent and his disciples. But the puzzle is that this was not established until the experiments of John Smeaton and Déparcieux were carried out about thirty years later.

---

* Swift's *Gulliver's Travels* (1726) contains also a remarkable prediction that Mars has two satellites and then goes on to give, with astonishing accuracy, the parameters of their orbits. The two satellites, Deinos and Phobos, were not in fact observed until late in the nineteenth century.

NEWCOMEN AND THE FIRE-ENGINE

The scientific bases for the invention of a successful fire-engine, using atmospheric pressure and the properties of steam as intermediary means of harnessing the power or effort of 'fire', had been laid by a succession of able continental scientists and engineers with, the assertive Savery notwithstanding, little or no British participation. The first stage, the solution in principle, had been achieved; the second stage, the solution in practice, was now realised through the genius of an Englishman Thomas Newcomen (1663–1727), and his assistant, John Cawley. Newcomen was born at Dartmouth in Devon; he was an ironmonger by trade and had some connections with the Cornish mining industry. A member of the Calvinist sect of Anabaptists, he died in London, probably of the plague, and is believed to have been buried in the Dissenters' cemetery at Bunhill Fields. Beyond these bare facts we know very little about Newcomen; of Cawley we know virtually nothing.

Francis Bacon would, one feels, have been delighted with Newcomen; not only because the invention of the first successful fire-engine confirmed the grand principle that the advance of natural knowledge must lead to improved opportunities for invention, but also because Newcomen was an Englishman; the first Englishman in fact to make a really major invention, comparable in importance with Gutenberg's printing press and the weight-driven clock.

The working of the Newcomen engine is easy to understand even if the theoretical principles underlying it are rather more subtle than they seem to be at first sight. Basically the engine consists of a large metal cylinder fitted with a piston which is connected by means of a chain to one end of a massive beam, pivoted in the middle and coupled at the other end to a mine pump. The cylinder is filled with steam from the boiler underneath and when the piston has reached the top of the cylinder, having been drawn up by the weight of the pump gear at the other end of the beam, the supply of steam is cut off, a jet of cold water is sprayed into the cylinder and the steam condenses leaving a vacuum under the piston. Atmospheric pressure then drives the piston down and when it reaches the bottom of the cylinder the condensing spray is turned off, the supply of steam is turned on again and the cycle recommences. The cold condensing water is supplied from a small cistern near the top of the engine which is kept topped up by means of a small force pump driven by the motion of the great beam. The used (and therefore hot) condensing water plus the condensed steam drain out of the cylinder through an 'eduction' pipe, fitted with a valve, and thence pass into the boiler.

The first Newcomen engine was very possibly put up in Cornwall in about 1705. Unfortunately no first-hand records have survived but almost certainly this first prototype machine would not have been automatic, or self-acting. That is to say, as with the Savery machine, the engine-man or his assistant would have had to have turned the supplies of steam and of condensing water on and off during each cycle. As the engine would have worked quite slowly this would not have been too difficult a task. But when the first engine of which we have a definite record—together with a drawing—was put up near Birmingham in 1712 automatic operation had been developed and the engine was entirely self-acting. The oscillations of the great beam switched on and off the water and steam at appropriate times in the cycle and the engine-man had nothing more to do than to tend the fire and keep a watchful eye on the working of his machine.

An inspection of the extremely ingenious automatic valve mechanism indicates clearly enough that a great deal of time and effort must have gone into devising and making it effective. The essential requirement is that the operation of turning on or off the steam or water must be carried out very quickly, with a snap action, for the engine cannot work if steam and condensing water are entering the cylinder at the same time. The motion of the great beam is necessarily slow on the working stroke because of the weight of water being lifted by the pump; and it is slow on the return stroke because before the piston can rise the steam has progressively to heat up the cylinder sufficiently so that condensation does not take place; this is a slow process for metal cylinders necessarily have a large heat capacity. The problem then is to transform the slow oscillations of the beam into snap actions shutting off or turning on the steam and water. Newcomen solved it by causing the great beam to trip what were, in effect, gravity operated valves so that the speed of switching on or off was substantially independent of the speed of the beam.

We can briefly summarise the novel features of this invention although we must remember that in this case the whole is greater, very much greater, than the parts:

(1) The separation of the boiler and cylinder allowed much more rapid action than would have been possible with Papin's suggested engine.

(2) The idea of using an internal spray of condensing water also enabled quicker condensation and therefore more rapid action.

(3) The valve mechanism represents a display of inventive ability amounting to genius. Our own experiences in building, to one-third scale, as accurate a replica as possible of the first Newcomen engine* suggest very strongly that the valve mechanism was not first developed on a small model but was almost certainly worked out by trial and error and built into a full scale working engine as experience was gained. It is, in fact, by no means easy to get the valve mechanism to work properly on a small model; even with all the wisdom of hindsight and the resources of a modern workshop.

(4) Water contains dissolved air and as the steam entering the cylinder is always accompanied by a certain amount of this air the engine would stop after a number of strokes, 'air logged', for no amount of cold water would condense the air. Newcomen got round this difficulty by fitting the cylinder with a 'snifting' valve through which the air could be flushed by the incoming steam once a cycle, thus to prevent the engine becoming air logged. Although the presence of air dissolved in water was fairly well known in Newcomen's time it does not require much imagination—or experience of development work—to realize that the diagnosis of 'air logging' and its elimination by means of the snifting valve must have caused Newcomen a great deal of trouble. It is, incidentally, significant

* The engine was built in the Department of Mechanical Engineering workshops at UMIST for the Manchester Museum of Science and Technology. The drawings were made by Dr C. T. G. Boucher and the work executed by Mr J. Flowett and Mr G. Needham. The engine stands about ten feet high and develops about one quarter horse power.

that Savery made no provision for expelling the air from his big condensing vessel; on the contrary he explicitly advised against flushing out the vessel with steam for that would be to waste it. One suspects either that the Savery engine never worked or that engine men were in the habit of ignoring Savery's advice and blowing the air out of the engine every so often.

(5) The Newcomen engine actually worked and competed successfully with the long established modes of power—wind, water, muscle. It was safe (the maximum pressure of steam employed was no more than a pound or two above atmospheric pressure), reliable and within the capacity of the technics of the time to build and to operate. It filled a long felt need in important branches of early industry and it was, as events showed, capable of unlimited development, being the first established heat engine. As such it belongs to the very select group of major inventions that have decisively changed the course of world history.

The Newcomen engine spread fairly rapidly. It was put to work draining the coal mines on the north-east coast of England and in the midlands and it provided a very welcome addition to the limited power resources available for pumping out the valuable non-ferrous metal mines of Cornwall. In the third decade of the eighteenth century it reached the important and technologically very progressive mining area of Schemnitz in Hungary (now Slovakia), as well as Sweden (Dannemora) and France (Passy). Engines were erected in Scotland and in later years in Spain, Russia and in South and North America.

The first published accounts of this engine were given by Jacob Leupold (*Theatrum machinarum generale*, 1724), Stephen Switzer (*Hydrostaticks and hydraulicks*, 1729), Martin Triewald (*On the Fire Engine*, 1734), Bernard Forêt de Belidor (*Architecture hydraulique*, 1737–1770) and J. T. Desaguliers (*A Course of Experimental Philosophy*, 1734 and 1744).

Leupold ascribed the invention unambiguously to Isaac Potter, the English engineer who erected the first atmospheric engine at Schemnitz. He made no mention of Newcomen. Belidor was more accurate in his account of the invention of the engine. After mentioning Papin's suggestions he remarks:

> . . . one cannot deny that M. Saveri [sic] made the first successful fire engine; this is confirmed by several letters which

were sent to me on that occasion by Fellows of the Royal Society of London: they also mention a Mr. *Newcomen* as having contributed much to the improvement of these machines.

The Fire Engine having many different components, it is convenient in order to avoid confusion to expound the principles so as to clarify the mutual relations of the parts. We see at once that the mechanisms of these machines basically depend on a beam, one end of which is connected to a set of aspirating pumps and the other to a piston moving in a cylinder.

The cylinder is connected to a big copper alembic [sic] and both are well sealed so that the outside air cannot enter. The bottom of the alembic forms the roof of a furnace the fire in which is the motor, or moving agent of the machine.

There are a number of interesting points about this short passage. In the first place we see that Newcomen is now given some credit for the invention but the essential differences between his engine and that of Savery are blurred over. In the second place we note once again the inadequacy of the current technical terms to describe a new machine: the boiler is referred to as an 'alembic',* which was the name for an alchemical vessel. On the other hand Belidor is fully aware that the operation of the engine depends on a few basic and quite general principles. Finally, and perhaps most interesting of all, there comes a clear and unambiguous recognition that the motive agent of the engine is fire.

It is more than likely that one of Belidor's London correspondents was the Rev. J. T. Desaguliers, FRS: in which case it would hardly be surprising that two of Belidor's observations were also made by Desaguliers:

Mr Newcomen, an Ironmonger, and John Cawley, a Glazier, both living in Dartmouth, Brought it (the fire engine) to the present Form in which it is now used and has been these thirty Years.

Tho' his Method differs much from Captain Savery's and the Force of the Engine is quite different; yet it is wrought by the same Power, viz. the Expansion of Water into Steam and that Power is raised by Fire.

Justice had at last been done to Newcomen, for Desaguliers' account was the best to appear in the eighteenth century and

---

* Later in the same passage Belidor uses the word 'chaudière' to denote the boiler.

it formed the basis for many other descriptions of the origin and nature of this most remarkable invention.

Perhaps the best way to appreciate the difficulties that must have beset Newcomen is to ponder Mr G. Needham's account of his experiences in getting the Newcomen engine in the Manchester Science Museum to work satisfactorily. This machine is probably the first sizeable one of its type to be built since the eighteenth century. It differs from the earliest known engines only in that steam is raised by means of immersion heaters and that the supporting structure is strengthened by steel girders. Mr Needham writes:*

The first test run was made almost exactly twelve months after we began the design and construction of the engine. Although restricted to manual control the first run went off very well indeed. But attaining satisfactory automatic control proved very difficult. An inverted injector nozzle remained undetected for three days and this was followed by an obstruction in the force pump valve chamber. There were a number of incidents like these; unfortunately as the engine reacted in much the same way for very different faults, each one could only be tackled by dismantling the major units one by one.

At this time the boiler was consuming some 8 kilo-watts. With various combinations of weights and nozzle sizes I was able to get between 10 and 20 full strokes before running out of steam. An additional 4 kilo-watts was required in order to maintain continuous motion at 8 to 10 strokes per minute; but this left little steam in reserve. Finally, the power was raised to 16 kilo-watts [equivalent to about 20 horse power] and with suitable adjustments the speed was increased, marginally, to about 12 to 14 strokes per minute.

A careful watch had to be kept on the water level in the cistern, for when the engine was running at its maximum speed the force pump delivered more water than was required and even the overflow pipe could not cope. In this case an adjustment had to be made to the by-pass valve on the pump. On the other hand if the engine ran too slowly, more water was needed than could be pumped into the cistern, the syphon was broken and the engine came to a halt.

---

* In a note to the author. Dr R. L. Hills and Mr Needham are preparing an account of their work and experiments on this engine which will be published in due course.

I found it necessary to maintain a two inch level of water on top of the piston at all times. This prevented air being drawn into the cylinder when there was a vacuum under the piston and it also assisted the cooling of the cylinder.

There is no difficulty in starting the engine from cold as the cylinder will readily condense steam for several strokes, thus allowing the cistern to be primed with water. This phase is best carried out manually, with the injection valve closed, until the cistern has begun to fill with water. It is, however, a different story when the cylinder is hot, the engine stopped and no condensing water available. In this case the injection valve must be shut and brief bursts of steam allowed into the cylinder so that the engine makes short strokes; in this way water can be pumped up to the cistern to permit normal operation.

Provided enough steam is available for the desired speed there are, I think, six factors that govern the performance of the engine. They are:

(1) Steam orifice size. This should be wide enough to allow a slight build-up of steam in the cylinder at the beginning of the upwards stroke so that the snifting valve lifts and the included air can escape. But the pressure should not be so great that the steam assists the piston in its ascent.

(2) Piston weights. Weights should be added to the piston so that the combined weight of the piston and rod equals the effort of the steam underneath.

(3) Mine pump weights. Weights should be added to the pump rod shaft to control the speed at which the piston rises and so to decide the time that the snifting valve stays open.

(4) Injector nozzle size. The injector nozzle must be of such a size that no more than sufficient water enters the cylinder to ensure condensation. If it is too wide then as the vacuum forms water will be sucked in in such quantities that the cylinder floods. The injection water should have enough velocity to form a jet that hits the piston and rebounds as spray.

(5) and (6) Snifting and eduction valve weights. These valves work as a pair and must be adjusted as such. In other words for any given speed the eduction valve must always be slightly heavier than the snifting valve; they will then be balanced in such a way that air can escape through the snifting valve while the heavier condensate escapes through the eduction valve. Too great a weight will cause air and water to accumulate while too little will allow steam to run to waste and may vitiate the vacuum.

THE ORIGINS OF MODERN TECHNOLOGY

A sense of history is a sense of civilisation, for without civilisation there can be no history; and, one might add, without some awareness of history there can be no civilisation. The development of the idea of history from early fables and legends about gods and heroes through the period when it was concerned exclusively with public affairs—military and political—up to the more subtle ideas of social evolution propounded by the French and Scottish *philosophes* of the eighteenth century has been slow and hesitant; which is perhaps not surprising when the inherent difficulties of the subject are remembered. Concurrently it has been very hard for historians to rid their minds of accumulated prejudices. Objectivity is a dearly bought intellectual commodity and what is unpleasant or distasteful or hurtful of national pride tends to be forgotten or distorted beyond hope of fair judgment.

Mining is a case in point. Most people would find working in a mine arduous, dangerous and excessively unpleasant at all times. Moreover in Britain the mining industry has been associated with a chronologically long and socially extensive record of disasters—individual and communal. It is hardly surprising therefore that mining rarely receives its full recognition for the major contributions that, as an industry, it has made to the development of European civilisation. Indeed, it is probably true to say that, more than any other, European material culture has been based on mining. This has been at once a cause and an effect of European technological superiority.

The mining areas of Europe in medieval, renaissance and modern times were centres of technology and science as well as of financial organisation and business enterprise.* Not only do a number of basic technologies and sciences converge in mining—chemistry, metallurgy, medicine, civil engineering,

* Among the technologists and scientists whose careers were associated in part or in whole with mining areas were Agricola, Ercker, Paracelsus, Savery, Newcomen, Smeaton, Watt, Trevithick, Humphry Davy, Werner, von Buch (geologists), the Fuggers and Welsers (bankers). The great mining academy at Schemnitz was probably the first technical college—or university—in the world.

geology and mechanics, hydraulics and transport—but mining generally is that part of industry that sets the really hard, the 'man-sized' problems; those of controlling forces much greater than those normally encountered in human technics and of effecting changes in materials on a much more extensive scale than craftsmen are accustomed to.

There can be few more instructive comparisons, offering deeper insight into the technology of a given age, than a juxtaposition, in imagination, of the Newcomen engine on the one hand with John Harrison's chronometer of roughly the same, mid-eighteenth century, period. The former, an example of which can be seen at South Kensington, is by almost any standards an ugly brute of a machine, raw and unfinished; the latter, which can be seen at the National Maritime Museum at Greenwich, is possibly one of the most elegant and pleasing machines ever made. Harrison's chronometer kept time so accurately that it enabled the longitude to be determined at sea by a comparison of local (sun) time with chronometer (Greenwich) time. This solved a very old problem and won a prize for its maker. The internal precision of the chronometer is reflected in its aesthetic appeal, but it was, of course, the end-product of nearly five hundred years of development by skilled horologists while the Newcomen engine was only at the very beginning of the new technology of the heat engine. By the nineteenth century heat, or rather steam-engines, had reached a chronometer-like standard of perfection and elegance. The same sort of evolutionary trend to perfection is apparent if we compare, say, Blenkinsop's clumsy locomotive of 1811 with 'Mallard' (one of the last of the great steam locomotives), or the Wright bi-plane with a modern jet-liner.

An enlightened observer of early eighteenth-century technology might well have concluded that the branch of industry with the most rosy future, technical and economic, would be mining. Mines and metals would surely form the sinews of future national prosperity. As for the great fire-engine of Newcomen that was not so much a revolutionary force in its own right as one more indication of the central importance of mining and metals: a gift from 'philosophy' to the mining industry. Such a judgment would be correct, but it would not be the whole truth. For very surprisingly one of the oldest and most basic crafts in the world—textiles—suddenly became a growth industry, perhaps the growth industry of the series of linked changes that constitute what is known as the industrial revolution. Our hypothetical observer, however, could hardly have foreseen this: the very idea of a 'growth' industry would have been unintelligible. Progress, for him, would have meant the piecemeal improvement of existing trades and practices along lines that were already established. Any appreciation of the possibilities of unrestricted, or as we may term it 'open-ended' technology, lay far in the future.

There are broadly speaking four main processes, or rather groups of processes in the fabrication of textiles. There is the preliminary cleaning and arranging or combing of the fibres, this is followed by the spinning of the fibres into a thread, then the weaving of the thread into fabric and finally there is the bleaching and dyeing of the woven fabric. During the course of the eighteenth century technologists, mainly English, succeeded in mechanising the first three processes and, thanks to the rapid advance in the science of chemistry towards the end of the century, greatly increasing the speed and efficiency of the fourth.

Traditionally England's wealth had depended on the wool trade—the symbolism of the 'woolsack' is well known—and the magnificent medieval abbeys of Rievaulx, Jervaulx, Bolton and Fountains were paid for by the profits from vast herds of sheep. But it was an export trade; the weaving and dyeing were done in the Low Countries; the English, as Defoe intimated, supplied only the raw material, not the technical 'know-how'. As for cotton fabrics, in the early years of the eighteenth century they were imported from India. The beginning of the process of mechanisation in the textile industry began when the half-brothers Thomas and John Lombe renovated a silk mill which had been set up by one Thomas Cotchett at Derby and which was powered by the river Derwent. The Lombes had, by means of ingenious industrial espionage, discovered the secrets of the silk-throwing—or spinning—machinery in use in northern Italy. Once again English technology was derivative, this time from those developed mechanical skills so character-istic of Italian genius and evidently related to artistic and architectural abilities.

However the silk industry is peculiar in two respects: it is essentially a luxury trade and production on the grand scale is hardly possible; furthermore the problems of spinning silk by machinery were relatively simple. The fibres are very long, sometimes up to several miles long, and being sticky they hold together without much difficulty when spun. The main difficulty indeed is the preliminary one of unwinding the silk from the cocoon. But with commoner textiles, like cotton and wool, the difficulties in the way of mechanisation were more acute. The fibres have to be prepared as in the case of silk spinning, but they are much shorter than silk fibres so that giving them the right amount of twist to form a strong, uniform thread is much more critical. From time immemorial the business of spinning had been the prerogative of women whose more sensitive and skilled fingers could draw out and twist the cotton or wool to give a fine thread. The medieval invention of the spinning wheel had increased productivity but had not diminished the need for skill on the part of the spinner or 'spinster'. To try to mechanise such a process using the clumsy wooden machinery of the eighteenth century would, on the face of it, have appeared to be an impossible ambition. Surely the fragile threads would snap, and keep on snapping? Yet it could be done for silk, so why not for wool and cotton? And there was now an added incentive for the mechanisation of spinning, if it could be done, for in 1733 John Kay invented the fly shuttle (or, as it came to be known, the flying shuttle), and as this increased the efficiency of weaving it necessarily increased the demand for thread. In a word spinning was now a serious bottleneck.

The loom is, of course, a very old instrument and the action of weaving is, in all essentials, very simple (Figure 19). A series of threads are drawn out parallel to one another and very close together. They are called warp threads and alternate ones are lifted up by means of vertical heddles so as to form a prism of triangular section called the 'shed'. A thread—the weft—is now passed along just inside the apex of the shed and at right angles to the warp. The positions of the warp threads are now interchanged. Those that were raised are lowered and the others are lifted up, thus binding or weaving in the weft. In this way a woven fabric is built up, the weft being regularly pressed together by means of a reed or comb whose teeth pass between the warp threads.

The weft is wound round a bobbin and is paid out as the bobbin is passed manually from side to side of the shed. This is necessarily a rather slow process and the width of fabric that can be woven is limited to the distance a man can reach to pass the bobbin through the shed. John Kay's invention greatly increased the speed of the operation and also the width of fabric that can be woven. For with the flying shuttle, as the name suggests, the bobbin, carried in a shuttle, pointed at both ends, is projected at high speed from one end of the shed to the

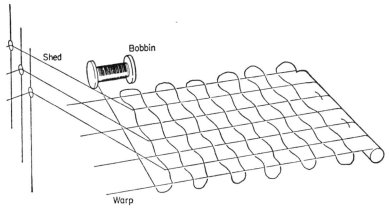

Figure 19.

other and back again. This is done by placing the shuttle on a horizontal 'launching pad' or shuttle box at one side of the loom so arranged that by jerking a cord the shuttle, which is fitted with small wheels or rollers, can be shot at high speed across the loom to the opposite shuttle box, and thence jerked back to the first. In this way the weaver does not have to leave his seat in front of the loom and the width of fabric he can weave is limited only by the distance the flying shuttle can travel.

The first significant attempts to mechanise spinning and so catch up with the demand of the weavers were made by John Wyatt and Lewis Paul in 1738 and in 1758. The spinning machines which they patented in those years showed strong similarity to the silk throwing machines in Lombe's mill. But one distinctive innovation they made was to use rollers (Figure 20). The cotton or wool was fed, in the form of a loose rope, or roving, through a pair of rollers and then passed to a flyer that twisted it and wound it on to a bobbin. The roving is supposed to be drawn out into a thread between the rollers and the flyer. But the device was never satisfactory and it was left to Richard Arkwright to perfect the process and make possible the successful spinning of cotton or wool by mechanical means.

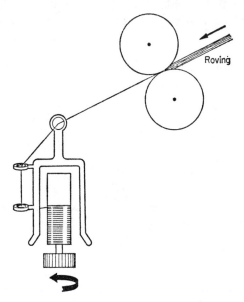

Roving

Figure 20.

The development of the textile industries in the eighteenth century made increasing demands on the limited power resources of Britain and so emphasised the urgent need for the utmost economy: squeezing the last ounce of duty out of the last drop of water falling the last inch. For it was basically on water-power that the textiles industries depended through the eighteenth century and for the first two decades of the nineteenth.

Parent's 4/27 rule had been accepted by engineers and mathematicians. Belidor had put his considerable authority behind it and had gone on to argue, by an egregiously wrong-headed piece of reasoning, that the efficiency of an overshot wheel could only be about one-sixth that of an undershot one and to urge that wherever possible overshot wheels should be converted to undershot operation. There were, however, those whose experiences tended to cast doubt on the validity of Parent's conclusions. Stephen Switzer, for instance, reported that overshot wheels were generally more efficient than undershot ones and Desaguliers observed:

> I have had occasion to examine many Overshot and Undershot Mills and generally found that an Overshot Mill ground as much Corn in the same time as an Undershot Mill with ten times less Water, supposing the Fall of Water at the Overshot to be 20 Feet, and at the Undershot to be about Six or Seven Feet. I generally observed that the Wheel of the Overshot Mill was 15 or 16 Feet Diameter, with a Head of Water of Four or Five Feet, to drive the Water into the Buckets with the same Momentum.

Most important of all, however, was John Smeaton (1724–1792), who concluded after very extensive experience designing, erecting and repairing water-wheels that undershot wheels were usually rather more efficient than 4/27 and that overshot wheels were considerably better. Moreover the wheels Smeaton was concerned with were not assumed to be perfect; they were actual working wheels, subject to all the usual imperfections of the real world. Smeaton therefore decided to put the matter to an experimental test. He made a model water-wheel, about two feet in diameter, that could be driven either overshot or undershot and that, by means of a pulley and axle, could wind up a scale pan in which different weights could be put. The apparatus was designed so that the head of water driving the wheel could be kept constant at a pre-determined level and due allowance was made for the friction losses when the wheel was turning.

As a result of an extensive and very careful series of experiments during which the load was varied from zero to that which stopped the wheel, Smeaton found that for different quantities of water flowing in the stream the maximum ratio of effect to natural effort amounted to about 1/3 rather than 4/27 for an undershot wheel, and about 2/3 for an overshot wheel.

While the undershot wheel was most efficient when its load allowed it to turn with about half the velocity of the stream the overshot wheel became more and more efficient as its speed became less;* which implied, of course, that it had to have bigger and bigger buckets. Perhaps this explains why Parent did not consider the overshot wheel: while it was quite obvious that the slower the wheel turned the greater the 'force' it exerted on the load—like a man struggling with a burden only just within his strength—it might seem that, on Galilean principles, slow working meant that little power, or 'effect', was being developed. In short, only actual experience (and experiment) with working water-wheels could lead to the conclusion that the overshot wheel, acting by weight and not impact, and suitably geared to raise a load at a convenient speed, was more efficient than the undershot wheel.

If, then, Parent was wrong in his conclusions, was it not reasonable to assume either that he had made a mistake (which in a sense he had) or that his premises were wrong? Smeaton did not go on to consider this larger issue. It was enough for him that the ratio 4/27 was wrong. He had no inclination to question the structure of theory and argument that had yielded this fraction; and we can hardly blame him for it. The theory had been shown by Colin Maclaurin to be entirely consistent with Sir Isaac Newton's mechanics and he would have been a rash fellow indeed who, in mid-eighteenth century England, had queried a Newtonian doctrine. Smeaton therefore recorded his experimental results meticulously, stated his conclusions clearly and left the matter at that. This was, in the circumstances, perhaps the best thing he could have done; in any case it won him the Copley Medal of the Royal Society (1759). But it left an awkward gap between theory and practical data that was to last, in England, for a further sixty years or so.

In his later papers Smeaton sought to explain the difference between the two cases. He inferred that the ultimate efficiency of an undershot wheel would be, ideally 1/2 and that of an

---

* A few years before Smeaton published his results Déparcieux had argued that an overshot wheel must be more efficient than an undershot one. His claim was based on the analogy that if two weights are connected by a string passing over a friction-free pulley the heavier weight will lift the lighter one, even when they are almost the same in magnitude.

overshot wheel, moving very slowly, would be 1. And the difference between the two efficiencies would, he suggested, be due to the waste of 'duty', as he termed it, or 'energy' as we should say now, in the forms of turbulence spray, whirlpools as the stream collides with the flat blades of the undershot wheel. This was a profound insight and a significant step towards the formulation of the concept of mechanical energy; but it was one that could only have been taken in the light of extensive practical experience. In terms of 'force' or Newtonian mechanics it was far from obvious that anything had been lost in the inelastic collision of water against blade: momentum, as all Newtonians held, is always conserved. In this respect the rival mechanics of Huygens and Leibnitz was much more satisfactory. Ideally the water should enter the water-wheel, or hydraulic engine, without wasting any effort, or energy, or *vis-viva* as the Leibnitzians had it, in generating turbulence or spray and it should leave the engine without any appreciable velocity having yielded up all its energy.

The full implications of Galileo's thoughts were now being unveiled. Engines should be designed in such a way as to exhaust all the energy from the moving agent, whether water or wind. There was something satisfyingly neat about the new insight; the most efficient transformations are those that take place without fuss or trouble! In the field of water power the effects were almost immediate. Wherever possible overshot water-wheels were erected in England. But if, for local reasons, an overshot wheel is not a practicable proposition the next best thing is a breast-wheel (Figure 21). This type of wheel goes some way to reconcile the simplicity and low prime cost of an undershot wheel with the doubled efficiency (2/3 compared with 1/3) of an overshot wheel. It was particularly suitable for slow moving rivers with little fall, and the type was used quite frequently to power the new cotton mills of Lancashire, Derbyshire and Cheshire.

Although the dogmatism of the Newtonians inhibited the development of a sound theory of water power in England things were different in France. In that country the authority of Newton was considerable, and rightly so, but it was not oppressive. Alternative theories of mechanics could be entertained by quite respectable writers; accordingly the *vis-viva* doctrine of Huygens and Leibnitz was formally invoked to explain the performance of different types of water-wheel. And in 1767 the Chevalier de Borda suggested that in order to minimise the loss of *vis-viva* (or energy) by the turbulent impact of water the water-wheel blades should be suitably curved to point upstream. So the theory of the turbine was born.

However we are concerned with the technological consequences of Smeaton's work and in this respect his influence was incalculable. He, his disciples and imitators, may be said to have increased significantly the efficiency of water power and this may well have been a factor in expediting the industrial revolution which was just beginning in England. From being

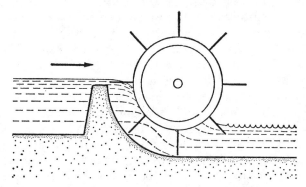

Figure 21.

a rather charming instrument of rustic economy the water-wheel became, in the closing decades of the eighteenth century, a 'science-based' and thoroughly industrial engine capable, in some cases, of developing several hundred horsepower to drive textile mills employing many workpeople.

Smeaton's contributions were not limited to improvements in water-wheels. He was a conscientious exponent of the new experimental technique (see *The Exemplary Experiment*, pp. 50–51) developed by Newton and other scientists of the seventeenth century. With his wide experience, his flair for mechanical efficiency and his deep knowledge of experimental science he soon came to see that the Newcomen engine of his time was much less efficient than it might be. He therefore made a model engine and carried out a series of systematic tests. One com-

ponent at a time was varied while the rest were kept constant.
As John Farey put it:

> Mr Smeaton, during four years, made a course of experi-
> ments, of which he recorded more than 130 in a book of tables,
> with calculations of the performance of the engine during each
> experiment; both with respect to the mechanical power exerted
> by the engine, and also the power it exerted during the time
> that it was worked by a bushel of coals. The experiments were
> in general continued until two or three bushels of coal were
> consumed.
>
> His practice was to adjust the engine to good working order,
> and then after making a careful observation of its performance
> in that state, some one circumstance was altered, in quantity
> or proportion, and then the effect of the engine was tried under
> such change; all the other circumstances except the one which
> was the object of the experiment, being kept as nearly as pos-
> sible unchanged.

In this way Smeaton found the dimensions and operating
conditions for the engine to yield the maximum power and
(what turned out to be the same thing) to work at the maximum
efficiency. He studied the best arrangements of the fire and
found, rather surprisingly, that for the greatest power and
efficiency, some of the steam in the cylinder should be left un-
condensed every cycle. This was because while total condensation would allow the atmosphere to exert its full weight (15 lb.
per square inch), cooling the cylinder right down to make a
perfect vacuum would slow up the engine so much and such a
great deal of steam would be required to warm it up again that
both power and economy would be greatly reduced. He found
too that a slight leakage of air into the cylinder improved the
performance as the air which could not be condensed by contact with the cool inside surface of the cylinder formed an
effective insulating layer between the working steam and cool
metal and so reduced the loss of steam due to excess condensation. The engine thus worked best on a mixture of steam and air
but, of course, the snifting valve was still essential to prevent
the engine ultimately becoming air-logged. As a result of these
painstaking studies Smeaton succeeded in roughly doubling the
efficiency of the Newcomen engine;* an order of improvement

* From about 5 million ft–lb. per bushel of coal to about 9 million ft–lb.
per bushel.

that was, curiously enough, about the same as the one he had made in the efficiency of water-wheels.

We can regard Smeaton's systematic method as the precursor of the familiar modern technique of evolutionary improvement that has led to the steady advance of such things as motor cars and jet air-liners. It is an exemplary method of getting the very best out of established machines but it cannot, or at least generally does not, lead to radically new or revolutionary improvement. For this one has, as it were, to frame new and bold hypotheses. To press the Newtonian analogy a little further: just as Newton had to make guesses about the nature of light when he had exhausted his systematic procedure in his inquiry into the 'celebrated phenomena of colours' so a fundamental improvement in the steam engine required new and drastic hypotheses about the natures of heat and steam. The Smeatonian method did not necessarily lead to this; the technique of experimental research did.

Smeaton's other activities included such things as the construction of harbours, bridges, lighthouses as well as the development of scientific instruments, and contributions to the sciences of astronomy and mechanics. His influence and that of his disciples was felt all over the country. Justifiably, for he was one of the first of a new breed: the professional engineer. We can regard the eighteenth-century civil engineer—the adjective distinguishes him from his military colleague—as springing from such established occupations as instrument and clock making, surveying, mill work and (according to Mr W. N. Slatcher) coal-viewing. Smeaton had been, like his great contemporary Watt, trained as an instrument maker but his experience spanned a very wide field indeed. In his latter years a Society, the 'Smeatonian Society', was formed in London. This society was the forerunner of the Institutions of Civil and, at one remove, Mechanical Engineers.

1769: ANNUS MIRABILIS

From an international point of view the most important year in British history since the arrival of Christianity was 1769, for in that year two most remarkable patents were lodged which, more succinctly than any other two events summarised, expressed and confirmed the industrial revolution. In January 1769 James Watt lodged his patent for the condensing steam engine and in May of the same year Richard Arkwright patented his water-frame for the mechanical spinning of cotton and wool.

As a young man James Watt (1738–1819) occupied a curiously privileged position for although he was a technician (an instrument maker) he was on friendly terms with a very select circle, including men like Joseph Black and Adam Smith, at Glasgow University. His approach to the problems of the Newcomen engine was necessarily academic and scientific rather than practical; this, as it turned out, was wholly beneficial and constituted a refutation of the old adage that an ounce of practice is worth a ton of theory.

Watt's analysis of the Newcomen engine began when he explained the dissimilarity in performance of a scale model and full-size engines. The first ran out of steam after a few strokes; the latter would work indefinitely. This he thought was due simply to the 'scale effect'. A small body, such as the volume of steam in a small cylinder, cools down more rapidly than a large one of identical form: such as the steam in a big cylinder. Hence the steam in the model cylinder condenses more rapidly than that in the big cylinder. This argument, while ingenious, was not original, for the great Dutch chemist and text-book writer Hermann Boerhaave had already propounded the theory of the scale effect on the heating and cooling of bodies of different sizes; and Watt can hardly have avoided knowing about Boerhaave's ideas.*

His attention having been drawn to the importance of the cylinder in limiting the performance of the Newcomen engine, Watt went on to measure the amount of steam supplied once a cycle to a working engine. He found that much more steam was consumed than would have been required to fill an empty

---

* He would have heard about them from Black if he had not read one of the three English editions of Boerhaave's great text-book on chemistry. As Mr R. J. Law has recently shown that the cylinder of Watt's model engine is not, as had been thought, to scale—its walls are relatively thicker than those of the full size engine—it seems more than likely that Watt's conclusion was based on *a priori* reasoning rather than on experimental facts.

D

space of the same volume as the cylinder. The inference was unavoidable: most of the steam must be used to heat up the relatively cold metal cylinder.

The obvious way to avoid this 'waste' was to make the cylinders of some substance, wood for example, with a much lower 'heat capacity' than metal.* But Watt soon found that while less steam was required to heat up and fill wooden cylinders than metal ones, if just enough water to condense the steam was injected there remained a considerable 'back pressure' in the cylinder that greatly impaired the efficiency of the engine. The only way to eliminate this back pressure was to inject enough water to cool the whole cylinder right down so that all vapour was condensed.

Watt knew that tepid water boils *in vacuo* for the phenomenon had been demonstrated by Professor William Cullen in Glasgow University. He now realised that it must be in principle impossible to have cylinders that stayed hot on the inside all the time and therefore needed to absorb little or no heat every cycle and at the same time to get a good vacuum by injecting only enough water to condense the steam. For in such hypothetical circumstances the condensate—the condensed steam plus the warm condensing water—must boil, thus vitiating the vacuum and generating back pressure. To harness the full pressure of the atmosphere one must have a good void and this entails cooling everything down so that there can be no steam or water vapour to oppose the descent of the piston. For economy, however, one must keep the cylinder hot all the time to cut down heat losses. These two conditions are plainly incompatible.

Watt had thus been brought by his experimental investigations to exactly the same point that John Smeaton had reached after extensive practical experience. Smeaton had been content to cut down heat losses by allowing an air leak and by retaining some steam in the cylinder all the time. The resulting opposition to the atmospheric pressure was more than compensated by increased fuel economy and more rapid working. But Watt was not so content: he was after perfection.

On a spring morning in 1765 Watt realised that the two basic

* A substance, that is, having low thermal conductivity and low volumetric specific heat.

conditions are incompatible only if one cylinder is used. If one has two the problem can be solved easily, for one cylinder, in which the piston moves, can be kept hot all the time, thus ensuring economy, while the other in which the steam is condensed can be kept cold and void all the time, thus harnessing the full pressure of the atmosphere. If the two cylinders are connected by a pipe, steam can pass from the hot, working cylinder to the other one (hereafter called the 'condenser') and will there be condensed: this leaves room for more steam that

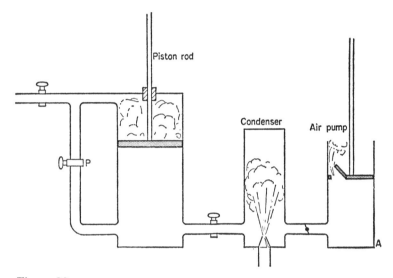

Figure 22.

rushes in to suffer the same fate. This Gadarene process continues until all the steam has left the hot cylinder and been condensed in the cold one, leaving a good vacuum in both. No time is lost in condensation, or in re-filling the hot cylinder with steam, and no heat is wasted in warming up cold metal. Economy, and the maximum power using the full pressure of the atmosphere have therefore been achieved.

There still remained, however, one source of heat loss to correct (Figure 22). If cold air is used to drive the piston down against the void, the hot walls of the cylinder will be unduly cooled. The consequential loss of heat can be eliminated if, instead of air, steam at atmospheric pressure is used to drive the piston. After it has done this duty the steam can be shunted under the piston, via the by-pass pipe P, and there condensed to form the void.

In this way James Watt invented the condensing steam engine. It was, of course, much more expensive and more complex than the simple Newcomen engine—it had for example to have an air pump in place of the snifting valve—but against this it was, as Watt's experiments showed, much more economical in fuel. This last factor made it a saleable proposition in those places where power was urgently needed and where economy of fuel was no less important. In particular it could hardly fail to appeal to mine-owners in places like Cornwall where there was no indigenous coal and power was urgently required to drain the mines.

This innovation could only have been made by a man of unusual scientific as well as technological abilities. A man, that is, who was familiar with the then known laws of heat and with a deep understanding of the properties of steam. It is profoundly unlikely that a practical engineer of the period could have been led to make this invention. On the other hand there still remains the old legend that Joseph Black, the discoverer of 'latent heat', told Watt about his discovery and on the basis of the realisation that 'there is a great deal of heat in steam' Watt proceeded to invent the condensing engine. The legend is untrue. Indeed it is in scientific terms nonsense: in terms of volumes there is *very little* heat in steam. But a detailed refutation is not really necessary: Watt himself was quite explicit:

> . . . [Professor Robison] . . . in the dedication to me of Dr. Black's 'Lectures upon Chemistry' goes to the length of supposing me to have professed to owe my improvements upon the steam engine to the instructions and information I had received from that gentleman, which certainly was a misapprehension; as, although I have always felt and acknowledged my obligations to him for the information I had received from his conversation, and particularly for the knowledge of the doctrine of latent heat, I never did, nor *could*, consider my improvements as originating in those communications.

This is clear enough, but Watt goes on to add that his inventions:

... proceeded solely on the old established fact that steam was condensed by the contact of cold bodies, and the later known one that water boiled *in vacuo* at heat below 100°(F), and consequently that a vacuum could not be obtained unless the cylinder and its contents were cooled (at) every stroke to below that heat.

The legend that Watt owes his improvement to the steam engine to Black's discovery of latent heat is therefore decisively refuted by the inventor himself. It is worth adding, however, that the greater the latent heat in steam the *less* the incentive to invent a separate condenser, for since the heat required to warm up the cylinder is independent of the properties of steam it follows that the greater the latent heat the higher the proportion of useful steam supplied per cycle.

Watt's patent specification was admirably clear. The principle of separate condensation is described and he goes on to claim that:

> I intend in many cases to employ the expansive force of steam to press on the piston or whatever may be used instead of them, in the same manner as the pressure of the atmosphere is now employed in common fire engines. In cases where cold water cannot be had in plenty, the engines may be wrought by the force of steam only, by discharging the steam into the open air, after it has done its office.

He concluded by outlining an engine that was intended to produce direct rotative motion, of the type that could be used to drive machines directly, instead of the familiar reciprocating action of the beam engine. In fact the final impression is that, apart from the Newcomen and Savery engines, Watt succeeded in patenting not only the atmospheric engine with condenser and his steam engine with condenser but also virtually every possible form of steam engine.

The story of Watt's partnership with Matthew Boulton and their eventually successful attempts to launch this complex and expensive new machine would take us too deeply into the fields of economic and social history to be repeated here. Boulton had quite exceptional vision and was possibly the only man of his time who foresaw clearly the coming age of steam power; the age that is in which steam engines would be used not only for pumping out mines but for a whole range of other purposes as well. Only such a remarkable vision could really have justified the risk that he, a successful Birmingham manufacturer, took when he entered into partnership with the poor, inexperienced but immensely gifted young laboratory assistant from Glasgow.

The possibility of using steam power to produce rotative motion had been grasped before even the Newcomen engine had been invented, but the difficulties and dangers of high pressure and therefore high temperature steam had effectively prevented anything being done. With the development of the viable low-pressure steam engine, however, the road was open again and Watt returned to the problem in his patents of 1782.

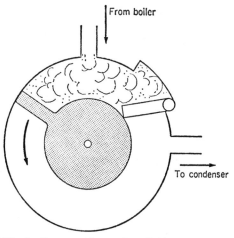

Figure 23.   *Watt's direct rotative engine, 1782.*

The rotative engine he then described was so simple that the reader coming across it for the first time marvels why such an idea was not immediately developed and why it did not supersede the clumsy beam engines of the period. Two concentric cylinders are set up, the inner one being fitted with a blade or piston, the outer one with a retractible baffle (Figure 23). Steam is applied between the baffle and the piston while on the other side of the baffle an exhaust port leads to the condenser. The inner drum, being moveable, will now rotate and the baffle must be retracted once a cycle to allow the piston to pass. There is therefore a portion of the cycle when no power is developed but this can be overcome either by fitting the engine with a fly-wheel or by coupling it to an identical engine whose 'dead space' occupies a different sector.

Watt's direct rotative engine is instructive. In principle it is surely the perfect solution to the problem of rotative power, the ideal engine. But, in practice, it did not work. The problem of retracting the baffle rapidly and smoothly enough to allow the piston to pass without crippling loss of power was never solved. Ideal though the engine seemed on paper, it proved, in practice, to be always just beyond reach. And although engineers of ability such as Hornblower, Maudslay, Bramah and many others were attracted by the idea they all failed to make it work. The problem of direct rotation was finally resolved by the invention of the high speed turbine of the 1880s. Very recently however the direct rotative Wankel engine, working on an internal combustion cycle, has been successfully developed by N.S.U. engineers in Germany.

Experience convinced Watt, as it was to convince others, that the beam engine was still the most suitable method of generating rotative motion. The increasing demands of a power-hungry economy acted as a strong stimulus on Watt to devise methods of converting the reciprocating action of beam engines into a rotative motion smooth enough to power the new textile mills. In achieving this end Watt revealed another side of his genius: that which enabled him to make empirical, non-science-based inventions, the analogues, in Baconian terms, of the printing press.

There were four such inventions. The double-acting engine, whereby the steam was applied alternately above and below the piston thus providing power all the way round the cycle; the parallel motion that enabled the double-acting engine to transmit a push as well as a pull to the end of the beam; the sun-and-planet gear, devised to get round a supposed patent covering the application of the crank to all fire-engines and, lastly, the governor whereby the centrifugal force of rotating weights raised a lever which controlled the throttle supplying steam to the engine. This last was an early example of automatic control mechanism applied to an engine.

The parallel motion is illustrative not only of Watt's inventive genius but of his concern with costs as a determining factor in innovation. Unless there is some means of steadying

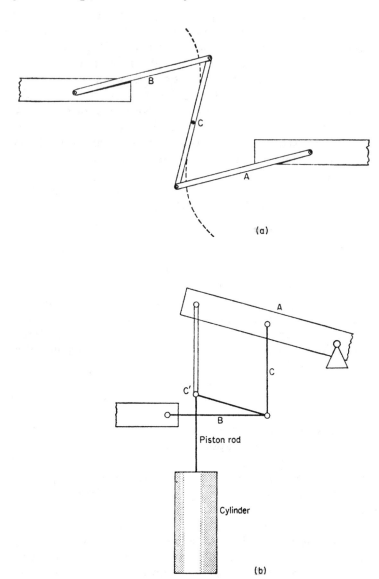

Figure 24.   (a) *Watt's parallel motion.*
              (b) *As finally applied to a beam engine.*

the piston rod when it is transmitting a push it will vibrate from side to side thus damaging the cylinder. It would be possible to prevent this by fitting the rod with a sliding cross-head moving along two parallel guide rods, as in the familiar steam locomotives of yesterday. But unfortunately in Watt's time the necessary machine tool, the planing machine, had not been invented. He had therefore to use his geometrical insights to solve the problem. Very probably his experience as a draughtsman helped him with this particular invention, the principles of which are set out in Figure 24. If the two arms A and B are of equal length and are hinged at opposite ends, then the centre of the third rod C will describe a straight line for an appreciable part of its travel. If, then, the piston rod is connected to the centre point of C it too must move on a straight line. Eventually Watt completed the parallelogram when he realised that the point C' will also describe a straight line. The parallel motion was one of the most important inventions in the development of knowledge about mechanical linkages.

The last of Watt's inventions that we shall discuss is that of expansive operation. It was an invention that was to have very important consequences and it reveals Watt's insight into the physical processes of power at its most acute. He explained the problem in a letter that he wrote to his friend William Small in 1769: 'I mentioned to you a method of still doubling the effect of the steam and that tolerably easy by using the power of steam rushing into a vacuum at present lost . . .'

Watt's solution is extremely simple. The supply of steam is cut off after the piston has travelled only a short way down the cylinder. The limited amount of steam above the piston continues to exert a pressure on it and to drive it down into the void space beneath. But as the piston moves so the pressure of the steam falls progressively,* until when it reaches the bottom of the cylinder the steam has expanded so much that there is little pressure, or 'spring' left. Nothing, in effect, has been wasted and in this way Watt hoped to extract the very last drop of 'duty' from the last puff of steam in his engine.

* It was assumed, not implausibly, that the pressure of the steam would fall in accordance with Boyle's law.

ARKWRIGHT AND HIS CONTEMPORARIES

Richard Arkwright (1732–1792) came from Preston in Lancashire. By trade he was a barber and wig-maker and he seems to have had little or no direct connection with the great engineers we have just been discussing. His first notable invention, that of the 'water-frame', was independent of scientific knowledge of his day and at first sight it appears to be little more than a development of Paul and Wyatt's spinning

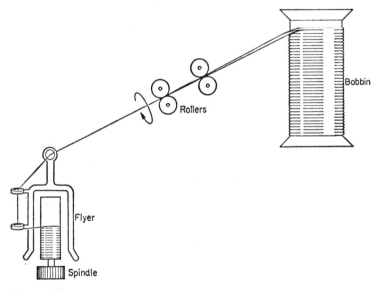

Figure 25.

machine—with this important difference however: Arkwright's machine actually worked. The cotton was first prepared, cleaned and carded (or combed) and then made up into rovings, or very loose ropes of cotton about a quarter of an inch in diameter. These rovings were wound on to a bobbin (Figure 25) from which they were fed through a first pair of rollers and thence through a second pair, rotating at a faster rate. These rollers, whose surfaces were made of fluted iron and of leather, drew out the rovings into a thread that was then twisted by a revolving flyer before being wound on to a spindle turning at a different speed. Arkwright's first water-frames, notwithstanding the name, were driven by horses. They could spin four threads at a time, but many more could be done if the power was increased so that additional sets of bobbins, rollers, flyers and spindles could be added.

Dr R. L. Hills points out that Arkwright succeeded where others had failed because he used *two* pairs of rollers revolving at different speeds and separated by a distance that was about the same as the length of the fibres to be spun. The other critical factors, Hills remarks, were the adjustment of the pressure between the rollers by means of suitable lead weights and the relative speeds of the flyers and spindles that had to be determined in order to give the right twist.

The water-frame, considered as an invention, hardly indicates the high genius of Watt's condensing steam-engine, but it was certainly competent enough. And, as Arkwright was to demonstrate, it marked the real beginning of the mechanisation of the textile industries: a revolutionary change in which he was to play several important roles, technological, entrepreneurial, social and political. Of necessity, perhaps, he was a tough character: domineering, single-minded and intensely energetic. He has had more than his fair share of denigration, but it is not too hard to detect the envy of lesser men in a good deal that has been said to his discredit.

As it happened, while Arkwright was working on his water-frame, another inventor was devising a different method of spinning that was later to be merged most successfully with Arkwright's machine. The second man's name was James Hargraves (or Hargreaves) of Stanhill, near Bolton, and his invention, patented in 1770, is known as the spinning Jenny. It is said that he got the idea when a spinning-wheel was accidentally knocked over: he noticed that the big wheel continued to rotate by inertia and thread could still be spun, for the now vertical spindle was clear of the floor. If this were so then there should be no difficulty in having a row of such vertical spindles and in drawing thread from each. The result was the first very crude Jenny. The rovings were wound on a row of bobbins from which they were paid out through a pair of parallel horizontal bars to a row of vertical rotating spindles (Figure 26). These spindles twisted the threads that were drawn out from the rovings as the bars were moved backwards. The bars were then clamped together, thus holding the threads firmly, while the backwards motion continued and the rotating spindles continued to give the requisite twist to the threads. Finally, a horizontal wire, called the 'faller', pressed the threads down the spindle so they could be wound on; as this was done the two bars were brought back to the starting position from which fresh threads could be drawn from the rovings. It is instructive to think of the two bars, alternately drawing and clamping, as acting in the same way as the thumb and forefinger would do

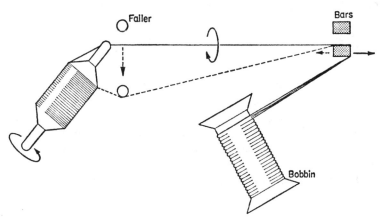

Figure 26.   *The mechanism of the Hargrave's Jenny.*

if a spinner was drawing several threads from a number of spindles at the same time.

The synthesis of Arkwright and Hargrave's machines was made when in about 1779 Samuel Crompton made the first 'mule'. The very name implies its hybrid nature and indicates why Crompton was unable to patent it. Arkwright's two sets of rollers were mounted at one end of a frame and they paid out the attenuated rovings to a carriage on which the spindles were carried (Figure 27). As the carriage moved back it drew out the threads that the spindles twisted. Before the carriage reached the end of its travel the rollers were stopped and gripped the threads, as the bars did in the Jenny, while the spindles continued to twist them. When the carriage reached the end of its travel the spindles were reversed for a few turns to free the threads which were then pressed down by the faller so that they could be wound on to the spindles. As the speeds of the rollers, the carriage and the spindles could all be varied independently of one another any desired type of thread could be spun on the mule.

In 1775 Arkwright patented a machine for carding cotton and a simple device using rollers for preparing rovings for the water-frame. Thus before 1780 the process of spinning had been mechanised. Essentially it had been broken down into a number of constituent processes and a suitable machine devised for each process in turn. The raw cotton was first opened out and cleansed of extraneous vegetable matter and dust by batting

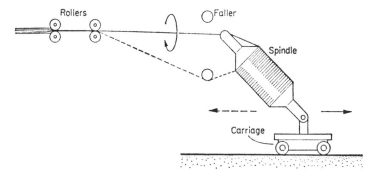

Figure 27.   *Crompton's 'mule'.*

and scutching machines; it was then carded, or combed so that the fibres were parallel to each other, and worked into rovings. These rovings were then spun mechanically by a water-frame, Jenny or mule into threads suitable for warp or weft. So great was the resulting efficiency, or productivity, of the spinning process that the bottleneck in the textile industries became weaving, and inventors' attention was turned to solving the problems of the power loom.

As early as June 1769, Arkwright had already entered into partnership with John Smalley and David Thornley, merchants of Nottingham, for the exploitation of the water-frame patent and the division of the profits to be expected; later on, Thornley dropped out, his place being taken by Jedediah Strutt and Samuel Need. In 1777 another agreement dissolved the partnership and vested all rights in Arkwright.

It was during these years, when the first cotton mills were put up by Arkwright and Jedediah Strutt, that the factory system was born. The very first water-frame had been a small machine, suitable for an industry that was still domestic. But once the way had been opened for large-scale production, then inevitably the economic rewards went to the largest enterprises. Rows of identical machines could be grouped together and arranged in sequence to cover all the separate processes from opening up the cotton to mule spinning and, ultimately, to weaving. No individual spinners or weavers could hope to compete with the new industry. Of course it was not merely a problem of inventing suitable textile machinery. A labour force had to be recruited, trained and provided for. Mill buildings had to be put up and suitable power sources tapped. Finally transport had to be arranged for the incoming raw cotton and the outgoing yarn and woven fabrics.

The early cotton mills were in country districts, often in very beautiful places, where rivers provided enough power for the great water-wheels that were required and where the local farming community could provide full or part-time employment for those members of mill-workers' families who were not employed at the mill. The early cotton masters, with enlightened self-interest, often provided good homes for their employees, putting up houses some of which are habitable today and will probably still be habitable for years to come. These early leaders were often Dissenters who were excluded from the fruits—some might say the corruptions—of office in State and Establishment. They were therefore free to devote themselves to business as their sole professional aim while the laws of England assured them their property and the profits their genius earned. In economic terms they may have had the optimum incentives with the best of opportunities.

The development of the rotative steam-engine finally enabled cotton mills to concentrate in towns, where labour was readily available and other facilities were close at hand. The steam-engine did not make the mills independent of rivers or streams for water was still required for boilers and condensers. But as the water was not needed to generate power, a head, or significant fall, was no longer necessary, and one mill did not have to monopolise a long stretch of river. Mills could therefore be crowded together on the same stream and their labour could be recruited indiscriminately from the impersonal town streets. This perhaps more than anything else was the cause of the later social disasters of the industrial revolution.

In the meantime, however, the construction of the new, large cotton mills necessitated a series of developments before the newly invented machines could be used effectively. One of the most important requirements was that the distance between the water-wheel on the one hand and the machines on the other should be kept to the very minimum in order to minimise the inevitable loss of power due to the friction. This implied a squarish, compact building with floors strong enough to stand the weights of the serried ranks of machinery and—here was the rub—fire-proof enough to survive. In those days when wood was, besides stone and brick, the main building material, when lubricating oils and fats ran on to wooden floors from bearings that might well run hot, or when naked oil lamps might easily be overset, a conflagration was always a distinct possibility. Firefighting appliances were rudimentary, hydraulic mains virtually non-existent and fire-proofing unknown; it was therefore hardly surprising that many mills were destroyed by fire.* It is well known that public buildings, such as the Theatres Royal at Covent Garden and Drury Lane were repeatedly destroyed by fire, but mills and industrial buildings in general have rarely enjoyed the affections of the public.

Among the first to concern himself with the structure of factories and with their fireproofing was William Strutt, FRS, the eldest son of Jedediah Strutt. William was a man of scientific attainments and a correspondent of some of the notable scientific men and engineers of Birmingham and the north-west. Iron was an obvious structural material for the technical progress of the eighteenth-century iron industry was now making it cheap and sound enough to use in buildings. In 1779 the world's very first iron bridge was built, across the Severn, near Coalbrookdale. Accordingly in 1792 Strutt began the erection of buildings at Milford and Derby in which vertical iron pillars were used and the floors of which were made of tiles supported on brick arches, instead of the usual wooden planks

---

* The most famous mill to be destroyed by fire was the Albion Flour Mill in London, burned down in 1791.

on joists. Later he continued in association with Charles Bage of Shrewsbury, to develop the complete iron-framed building. Galileo's theories on the strength of materials had filtered through and had come to the attention of scientifically minded mill owners. The process had perhaps been slow, but where, before the industrial revolution, had been the need for strong, multi-story fireproof buildings? In this way the early cotton masters could build mills which were reasonably fireproof and that were at the same time big enough and strong enough to contain large machinery and to stand up to the continuous vibration of its use.

Problems of structures, while fundamental, were not the only difficulties posed by emergent industry. The problem of transmission of power meant the development of the system of 'line shafting'. Horizontal rods, running just under the ceilings of each story carried the power from a vertical shaft, driven by the water-wheel, to the individual machines; the last link in transmission was the belting that connected the pulleys on the rods, or line shafting, to pulleys on the machines. Other problems included the heating and ventilating of the new buildings and these involved the application of the new (eighteenth century) science of heat. It was at this time that central heating, properly so called, began. If, today, we still miscall convectors 'radiators' that is merely a reflection of the fact that in those days the modes of propagation of heat—conduction, convection, radiation—were not yet clearly distinguished.

We may summarise these important developments very briefly. The invention of textile machinery by men like Arkwright, Hargraves, Coniah Wood, Crompton and many others was totally independent of science. It is true that the mechanical philosophy may have had some very general and indirect influence on the process of invention in the textile field but to all intents and purposes these inventions belong to the category that we have designated empirical and non-science based. But once the process became established, once the great new cotton mills began to spring up, a host of new problems arose that stimulated scientific inquiry and that were to give rise to new forms of industry. Thus by a process of chain reaction, new technologies were established that could hardly have been foreseen at the outset.

For the present it is surely enough to note that Arkwright became Sir Richard Arkwright and High Sheriff of Lancashire, but he never became a Fellow of the Royal Society.

A FOOTNOTE

The foundation of the science of heat and the development of thermometry in the eighteenth century had other implications for technology besides making possible the radical improvement of the steam-engine in its various forms. Following the establishment of an acceptable scale of temperature by D. G. Fahrenheit early in the eighteenth century it became possible accurately to measure the thermal expansion of metals and other solids. This enabled the instrument makers, led by men like Smeaton and Ramsden, to allow for or (better) to compensate for thermal effects. The 'compensated pendulum' is a familiar instance of the new technique. Thus a new standard of instrumental accuracy became possible, thanks to the new science of heat.

Chapter 4

The Years of Revolution, 1790–1825

The logic of industrial growth meant for the technologist and inventor that there was continuous pressure to innovate and to improve. Let us compare, in imagination, the attitudes of Dean Swift's Laputan land-owner with his domestic cornmill with that of, let us say, Samuel Greg who came over from Belfast and set up a cotton mill (which still stands) on the river Bollin near the picturesque village of Styal, just south of Manchester airport. At first the water-wheel was powerful enough to drive the mill but as Greg prospered he added to his building, installed new machinery and so increased his power requirements. A distinguished engineer, Peter Ewart, was brought in and the system was overhauled and improved. A dam was built, with new penstock and waterwheel so that a greater fall, or head of water, was obtained. This was sufficient until continued expansion again outran the capacity of the system; at this stage Greg had to instal a Boulton and Watt rotative steam-engine to boost the power of the water-wheel, when during summer droughts the level of the Bollin was too low to work the mill properly.

Economic growth meant in fact that the best available sites were soon snapped up so that newcomers had to develop the less good sites. Thereafter expansion of the markets would mean expansion of production and hence an increasing premium on efficiency of power generation and transmission followed by the application of a new source of power—the steam engine—to enable the expansion to continue.

It would of course be wrong to say that there was a 'power shortage' since there were then no national sources of power, as there are to-day. But it would be true to say that in the areas where the new industries were developing rapidly power was becoming an increasingly critical factor in expansion and more and more attention had to be paid to it. On the local scale therefore there was a power problem. Dr Hills puts the point very clearly: '. . . in 1788 there were about twenty mills in the whole of Oldham parish, eleven of them in Oldham itself, while by 1791 there were eighteen in Oldham alone. Because the best water sites had been taken most of these mills were driven by horses'.

Horses were by no means the only alternative source of power to be exploited in the anxious search for additional energy. It is recorded that in one mill a Newfoundland dog was set to work; the clumsy Newcomen engine was adapted to give rotative motion to drive Manchester cotton mills and even the Savery engine was revived in an improved and workable form. Necessity was certainly the mother of invention at this stage.

The prime source of power continued to be the water-wheel, however. This was mainly because a group of able engineers and technicians continued to improve its performance until large and highly efficient machines could be made. The development of structural engineering together with improvements in iron were contributory factors. Dr Pacey and his students have pointed out that a radical change took place at this time, comparable in importance to that which followed Smeaton's experimental paper of 1759. The use of light iron spokes *in tension* coupled with the technique of taking the drive, not from the axle, but from the rim of the wheel enabled much lighter and more efficient structures to be built. Indeed, in the case of these 'suspension' wheels the only strain on the wheel when it was working was its own weight. The ultimate refinement was depicted in the drawing of a water-wheel in Rees' *Cyclopaedia* of 1819. It represents one of Thomas Hewes' water-wheels. The utmost precautions are taken to prevent any loss of energy through the turbulent impact of the water on the buckets of the wheel, but now there is an added sophistication. A governor on the main shafting controls, through an auxiliary shuttle, the water passing on to the wheel. If the machinery runs too slowly the shuttle is opened and the wheel develops more power; if it runs too fast the shuttle closes and the power is reduced. An automatic control, or feed-back mechanism was in this way incorporated into the water-driven cotton mills. Plainly, the industrial water-wheel was very different from the rustic water-wheels of the early eighteenth century.

Indeed the water-wheel had reached a point of perfection beyond which only detailed improvements would be possible. Whether the stream was fast or slow running, large or small, a suitable water-wheel of overshot or breast form was available to suit most purposes.*

There remained, however, one limitation to the use of water-wheels: the maximum height of fall or head of water that could be usefully harnessed was no greater than the diameter of the wheel which would amount in practice to about forty feet or less. If one wanted to harness a stream of water falling several hundred feet the only way in which it could be done was by setting up a series of wheels, one above the other, in cascade. But this was obviously inefficient. A much better way of harnessing the power of really big falls of water was by means of column-of-water engines. These machines, which were developed from about 1740 onwards in South Germany and France, were like Newcomen engines but they were driven by the pressure of water and not that of the atmosphere. Later, as the Newcomen engine was progressively improved and gave rise to the Watt steam-engine so the analogous water-engine shared in the refinements and became progressively more efficient. Doubling-acting machines were made and the parallel motion was incorporated. There was even an analogue for 'expansive operation'. The supply of water was progressively throttled back so that the pressure on the piston diminished as it approached the end of its travel. The trouble with this method was that the motion of the piston became slower and slower as the water was throttled back, once every cycle. This did not matter if the engine was being used as a pump but it made it virtually impossible to generate rotary motion this way. Accordingly an alternative method was used: an air reservoir was attached to the high pressure water supply

---

* If one took Borda's advice, as Poncelet did in 1825, and curved the blades so that they pointed upstream, an undershot wheel could be made reasonably efficient since little energy would be wasted in the form of turbulence.

and this air provided the necessary elasticity once the water was throttled back. In this way the analogue of expansive operation could be applied to a rotative column-of-water engine.

The column-of-water engine was not much used in England although some leading engineers, like Smeaton and Trevithick, were greatly interested. It was mainly used on the Continent where falls of water were often much higher than in England, and supplies of good coal were less readily available. In any case it seems to have aroused a great deal of interest, possibly out of proportion to the numbers actually used. This was because it practically completed the full range of hydraulic power machinery. Engineers could now harness the power of a stream of water no matter whether it was big or small, fast or slow, high or low. All one needed to do with a stream of water falling several hundred feet was to enclose it in a suitable pipe and feed it to a pressure engine at the bottom of the valley. If the engine and its valves were strong enough and well designed then little or no energy need be lost in turbulent impact and the water could leave the machine without much velocity.

Thus by the opening years of the century it could be said that command over water-power was practically complete in both theory and practice. It required only the invention of the water turbine by Burdin (1824) and its manufacture to finish the art. Water-power technology, while much older than steam power, was indebted to the latter for many practical points, especially concerning the column-of-water engine; but in terms of theory water-power was far ahead of steam power. As the two technologies came together, the same sorts of machines being driven by steam and water, and as the same engineers at that time frequently dealt with both types it was perhaps natural that the 'cross-fertilisation' between the two technologies should not be all one way. In due course, as we shall shortly see, water-power technology repaid the debt to steam by suggesting an appropriate theory by which the operation of all steam-, indeed all heat-engines, became intelligible.

Parallel with the development of the water pressure engine was that of the high-pressure steam-engine. James Watt had set his face firmly, and in the circumstances not unreasonably, against the use of high-pressure steam: the engineering standards and the metallurgy of his time were incapable of dealing with such a dangerous commodity. Watt's extensive patents covered the heart of steam-power technology and it followed that engineers who did not share Watt's views about high-pressure steam could not put their ideas into practice. Had they tried they would at some point or other come up against one of Watt's key patents: the parallel motion, the sun-and-planet gear or the separate condenser. In 1800 however Watt's monopoly ended when his patents lapsed. The field was now open and engineers like Trevithick were not slow off the mark in developing a high-pressure steam-engine. Such an engine would offer many—and exciting—advantages. The higher the pressure the smaller the piston need be to develop the same power; and the smaller the piston and cylinder the smaller the whole engine. A really small powerful steam-engine might be put inside the hull of a boat to propel it by means of paddle wheels or screw; or the engine might be put on four wheels and coupled up to drive two or more wheels thus constituting a new 'horseless carriage'. In mining areas such machines could be designed to run on rails, hauling coal trucks.

The prospects, after 1800, must have seemed virtually unlimited. The steam, or the heat engine, had at least been freed from the restrictions imposed by unwieldy bulk and vast new markets were opened up. In addition it offered prospects of increased economy. Expansive operation was more readily applied to high-pressure engines: just a squirt of really high-pressure steam used expansively was obviously an economical way to work an engine. Furthermore the high-pressure engine was in terms of size and weight much smaller than the low-pressure one; and it should be much easier to guard against heat losses by carefully lagging the cylinder and steam pipes. *A priori* then the high-pressure engine should be more economical than the low-pressure one; and this was soon found to be the case.

It would be a mistake to assume that the history of the steam-engine at this time represented a straightforward and triumphant development of the standard reciprocating form: cylinder, piston, connecting rod and working beam. In fact the triumph of this type of engine was the outcome of a long and sustained struggle for survival among the many varieties of heat-engine that inventors proposed. A study of the patent rolls of that time indicates that a veritable morphology of engines could be drawn up of which only one general type, the piston engine, survived. It would be unjust to the many inventors concerned as well as a distortion of history to suggest or imply that we need consider only one type. Our morphology would include the following 'families' of heat-engines:

(1) Simple reaction engines. A jet of steam coming out of a horizontally pivoted pipe and directed at right angles to the axis of the pipe causes rotation in the same way that a simple lawn sprinkler works. Hero of Alexandria suggested such an engine and inventors never lost sight of the principle although they were never successful.

(2) A simple steam 'wind' mill. A jet of steam is directed against fan blades and causes them to rotate. Once again while very simple and obvious such engines were not successful. The speed of rotation would have had to be unattainably high, for those days, for the engine to have been reasonably economical and powerful.

(3) 'Buoyancy' engines. Bubbles of steam rising through boiling water, or of warm expanded air rising through a warm liquid, are trapped in the buckets of a kind of water-wheel immersed up to its axle in the liquid. The wheel therefore rotates. In the case of air bubbles the gain in buoyancy caused by expansion in the warm liquid should provide enough power to operate a pump submerging the cold air, to overcome the friction of the machine and then to leave a useful margin for outside application. The machine, which was first invented in 1792, was clumsy and inefficient but did offer a way of using waste heat; a common feature of many industrial processes.

(4) Direct rotative engines, of the type pioneered by Watt (see Figure 23, p. 90). These continued to be popular with inventors but were never successful.

(5) Steam-engines using piston and cylinder. These can now be conveniently subdivided into two classes:

1 *The high-pressure steam-engine.* Small, compact and powerful, it does not even need a condenser for, if the working pressure is

several atmospheres, the loss of one atmosphere's useful pressure is more than compensated for by the gain in simplicity and mobility. These engines enabled the development of the great railway systems of the early nineteenth century.

2. *Large stationary engines* in the (almost) traditional style. These would continue for a long time to be mainly beam engines; they might operate at moderately high pressures, would use condensers and, with skilled design and without restrictions as to space and size could be made very efficient indeed. Engines of this type continued to pump out mines and work cotton mills until well into the nineteenth century and in a few instances indeed almost up to the present. They represented the very pinnacle of beam engine perfection.

(6) There were also attempts, never successful, to harness the expansive force of heated metals and liquids. The fact that such expansions seem irresistible is entirely illusory for as they take place very, very slowly, extremely high ratio gearing is required to make them effective and with such high gearing the expansion is no longer 'irresistible'.

(7) Finally there were sustained attempts to find different and better 'working substances' than steam or water. Readily vaporisable liquids like alcohol and turpentine seemed, at first sight, to offer certain advantages but practice showed that that there was no gain at all in economy. Permanent gases like carbon dioxide, nitrogen and air were tried but none seemed better and all lacked the great advantage of steam. A very small volume of water converted into steam at a reasonably low temperature (212°F, 100°C) gives an immense increase in volume: something like 1,700 fold, or if expansion cannot take place, an immense increase in pressure. In the latter case the temperature of boiling must be raised above 100°C in order to vaporise all the water.

It was this very useful property of water, a universally available liquid, that made the success of the steam-engine possible. Or, to be more precise it was the first of the two related properties—the great increase in volume on boiling—that enabled feasible steam engines to be built long before there was any hope of engineers and metallurgists dealing with really high pressures and temperatures. This was not explicitly recognised and formally stated until 1824.

One of the engineers who felt the weight of Watt's (near) monopoly of patents was the able Cornishman Jonathan Hornblower. He had invented and patented an engine that incorporated the principle of expansive operation in a novel way. After the steam had driven the piston down in a small cylinder it was allowed to expand into a much larger low-pressure cylinder. This two-cylinder engine, in which the steam acted in two steps, first at high pressure then at low, marked the invention of the principle of compounding. Unfortunately when Hornblower tried to get his patent extended he fell foul of Boulton and Watt and lost his case in the courts.

In the course of the trial, however, Davies Gilbert, a Cornish MP, a mathematician and later President of the Royal Society,* gave evidence on Hornblower's behalf. Part of this evidence consisted of a mathematical analysis of expansive operation, and this came to be of some importance.

If we want to know the power developed by a steam-engine, then we must measure the pressure of steam in the cylinder, the area of cross-section of the piston and the length of the stroke and we must count the number of strokes per minute. The product of all these is the power expressed in ft. lb. per minute, or if we divide by 33,000, in horsepower.

If, however, the engine is working expansively then the pressure of steam in the cylinder varies and the actual power must be less than that given by this simple computation. It was important for men like Boulton and Watt to know the power being developed by an engine, for they sold their engines on the basis of the power they generated. We can measure the power of an expansively operated pumping engine easily enough; it is equal to the weight of water multiplied by the distance it is raised per minute. But if we have a rotative engine then the problem is much more elusive: we are not raising weights, we are turning mules and carding engines, power loom and other textile machines. Fortunately, Gilbert's analysis suggested a way in which the power could be measured and the cotton masters asked to pay a fair and agreeable price for their expansively operated engines.

* He was patron not only of Hornblower but also of Humphry Davy, Trevithick, Arthur Woolf and other members of the Cornish group of engineers and scientists.

Gilbert showed, and others confirmed, that if we plot a graph of the pressure of steam in the cylinder against the volume behind the piston then the area of that graph will be proportional to the work done in one cycle; if we multiply the work by the number of cycles per minute we get the power. The next step was taken by John Southern, Watt's able assistant, in 1796. Southern arranged for the engine to draw its own 'power curve', or as it came to be known, 'indicator

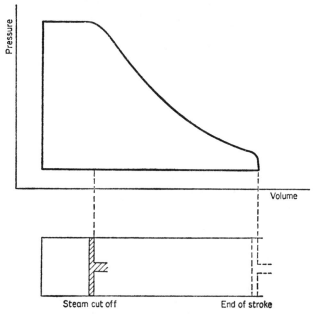

Figure 28.

diagram'. It was a very simple device: a sheet of squared paper is mounted so that one edge moves with the to and fro motion of the piston in the cylinder, and therefore with the increase and decrease of volume behind the piston (Figure 28). A pen or pencil held firmly in a clamp moves at right angles to the first direction of motion and presses on the paper. The clamp is moved by a small, spring loaded piston moving up and down in a small auxiliary cylinder connected to the main cylinder. The motion of the pen, then, measures the variation of pressure in the working cylinder and the combination of these two motions gives us the desired 'indicator diagram' from which we can deduce the power being generated by the engine and the way in which it is performing. The practical importance of this invention can hardly be exaggerated; but it also had some theoretical consequences, when Carnot and Clapeyron came to establish the science, as distinct from the technology, of thermodynamics.

Historians of science and of technology are always subject to the temptation to find in the periods they study unique factors that suggest it is a period of transition, a bridge, from one epoch to another. But reflection soon confirms that the effect is illusory, it comes from deep acquaintance with one particular period that necessarily highlights the differences between it and the ones that precede and follow. Another danger confronting the historian is, in effect, of the opposite nature. He may be tempted to iron out the differences between epochs; to see science and technology as a uniformly successful progression, much the same in one age as in the next. Save, of course, in those periods when (unaccountably) progress flags or stops, as in the dark ages.

The historian in fact must commit himself, consciously or unconsciously to some philosophy of the history of his subject. In the writer's view, then, there have been definite periods when decisive changes have taken place in the course of technological history. There was, for example, the medieval epoch of heroic, precocious inventiveness. Then, in the seventeenth century a highly articulated mechanical philosophy was placed at the centre of man's attitude to nature and mastery over it. A biological or organic approach to nature might quite possibly have led to a reasonably viable technology, more in sympathy with Aristotelean ideas than the one that actually triumphed. Such a hypothesis cannot readily be refuted for it was not until after the eighteenth century that the mechanical philosophy *per se* came to be closely associated with the progress of technology. Parent and Belidor, Smeaton and Watt ushered in the new technology.

Another period of decisive change appears to be the opening years of the nineteenth century. Many of the major technologies of the nineteenth and twentieth centuries were founded at this time and, one might add, the distinctive tone of nineteenth-century physical science, so decisively (if unconsciously) non-Newtonian, was set by men like Dalton and Young, Wollaston and Faraday. At the same time social changes took place in the organisation of technology and science setting them on the courses that led to the modern technological society. This is not entirely the wisdom of hindsight on the part of the writer. At least one contemporary observer noticed what was going on and drew the correct conclusions. 'We', wrote William Jackson in 1798, 'are doing the labour by which the Golden Age will profit'. Whether we regard our own age as characterised by the gold, or whether plutonium might not be the more appropriate element, is beside the point: our positive achievements spring very much from the labours of pioneers in the years of revolution 1790–1825.

The railway is a case in point. It was in 1804 that Richard Trevithick made the first steam locomotive run on rails and draw a line of trucks. This was at Samuel Homfray's ironwork at Pen-y-darran in South Wales. In 1811 Blenkinsop's locomotive started work on the Leeds–Middleton railway line, drawing coal trucks. It was a crude affair. Men had not even realised that the friction of the locomotive wheels on the track would be sufficient to pull the train: Blenkinsop's locomotive hauled itself along by means of a cog wheel that engaged a rack running alongside the track. The trucks were simply coupled together by links of chain and the driver walked alongside his charge exactly as he would have done had it been an extraordinarily powerful horse. Nevertheless it was, for its time, impressive enough.

> Shortly after our arrival we went to the coal wharf to see the arrival of the coal wagons which are set in motion by steam machines and which bring coals from the mines at a distance of six English miles from the wharf. It is a curious spectacle to see a number of columns of smoke winding their way through the countryside. As they approach we see them more and more distinctly until at length along with a column of smoke we also perceive the wagon from which it ascends, dragging a long train of similar wagons behind it, which gives it the appearance of a monstrous serpent.

Thus a German traveller in Leeds in the early nineteenth century (1816), and it is interesting to note that there was then no separate word for locomotive: it was a wagon, similar to all the ones it drew. But thereafter things changed rapidly. The extension of the railways began in the 1820s: passenger wagons, or coaches, were built and one by one a host of ancillary inventions were made whereby the great Victorian railway system became safe, efficient, comfortable and cheap.

At the time when the first steam railways were being planned, revolutionary proposals for flying machines were being put forward. Sir George Cayley, a wealthy Yorkshire landowner and a technological thinker—it would be misleading to describe him as an inventor—was a man who had both the intellectual capacity to work out the basic requirements for aeroplanes and the financial independence to do so without having to look for an immediate cash return on his labours.

The lighter-than-air machine, the balloon, had been invented as a direct result of the discovery of the atmosphere and, more immediately, of progress in the chemistry and physics of gases towards the end of the eighteenth century. By the early 1800s a number of daring ascents had been made, mainly by Frenchmen; the parachute had been invented and, impressively enough, demonstrated.* For all this the balloon was not a really satisfactory aircraft: it was not *dirigible* but merely floated about at the mercy of the winds.

Cayley, on the other hand, proposed in outline a feasible heavier-than-air and dirigible flying machine. He saw that the solution to the problem of heavier-than-air flight lay not in devising a set of flapping wings, a preconception that had misled all would-be inventors of flying machines before Cayley's time, but in having a fixed surface or plane: in fact that 'the whole problem is confined within these limits, viz: to make a surface support a given weight by the application of power to the air'.

This solution could only have occurred to a man who was

* The parachute was not first conceived of as a simple device for checking a fall by hindering the flow of air past the falling body. On the contrary it was explicitly realised that if a man's body (say) could in some way be harnessed to a vast bulk of air the specific gravity of the whole would be not much greater than that of air and the combined unit, man plus harness plus air, would fall through the surrounding air very slowly, like an egg through water. The parachute is therefore a device for effectively reducing the specific gravity of a man to something like that of air.

familiar with the properties of air and who, moreover, could reason scientifically from this knowledge. In much the same way Watt's separate condenser was invented by a man who was familiar with the properties of steam and knew exactly and in detail how it would behave under the given conditions. While, however, Cayley had reached or almost reached the stage of solution-in-principle he was still a very long way from devising a practical aeroplane. For one thing a suitable source of power was needed. Here the recent advances in the steam engine and the greatly enhanced interest in heat engines in general were relevant. What was needed was an engine that was as powerful as, but much lighter than, those which were being developed for steamships and locomotives. The conventional steam engine, with its water and heavy boiler seemed very unlikely to meet this exacting requirement. The air engine looked much more suitable.

Cayley's air engine consisted of a small pump that drove cold air into a furnace or combustion chamber in which it was heated so that the pressure increased thus driving a piston in a large working cylinder. The motion of this piston provided enough power to work the pump and to leave a useful surplus that could be applied externally. In conception this engine was not particularly original. The idea of using air as a motive or working substance can be traced back to a suggestion of Guillaume Amontons, put forward in 1699, while the incorporation of a pump to feed air to the furnace or hot body had been a feature of the 'buoyancy' engine and, much more recently, of the 'pyreolophorus' invented by the Niepce cousins. The latter was an air-engine that worked on the principle of internal combustion, grains of an inflammable substance being carried into a combustion chamber by the stream of pumped air and then ignited so that the resulting expansion drove out a large piston. The 'pyreolophorus' was reported to have propelled a barge upstream on the river Saône, although the story probably lost little in the telling. Cayley's engine was not very successful for, in fact, a great deal of development had to be done and much fundamental theory to be worked out before the air engine became a really practicable proposition. Nevertheless Cayley deserves great credit for establishing that the form of heat engine suited for the heavier-than-air flying machine must be the air engine.

The last of Cayley's suggestions that we shall mention was also related to his conception of a heavier-than-air flying machine. This was his proposal for a very light wheel, that would also be strong enough to serve as an aircraft under-carriage. The wheel was to be of the suspension type and in place of wooden spokes light, strong cords were to be used. This idea represented an important extension of the suspension wheel principle developed by such water-wheel engineers as Thomas Hewes and it is interesting to speculate how Cayley came to learn of it.

The attainment of Cayley's ambition lay some ninety or a hundred years in the future. Less dramatic but, as events showed, no less important than these pioneering efforts at land, sea and air transport, was the early development at this time of the machine tool industry. Machine tools have existed from time immemorial, the pole lathe being a very ancient instrument. But up to the eighteenth century lathes were used only for turning soft metal, woods and substances like ivory or bone. The screw-cutting lathe, employing a suitable lead screw, had been invented in the late medieval period but the screw threads it cut were in wood and were intended for things like wine presses and printing presses. The main use of the lathe up to the end of the eighteenth century had been, apart from turning wood, for ornamental work such as turning intricate patterns on metal (rose engines), for watch and clock making and for instrument making generally. These machines did not constitute, as Professor Woodbury puts it, 'industrial lathes'.

A form of machine tool that could claim—in terms of size, at any rate—to be 'industrial' and at the same time to be of respectable antiquity was the boring machine. These were developed for producing uniform bores in cannon castings and they are depicted and described in Biringuccio's *De la piro-technia* (1540). The casting was secured horizontally and a cutting tool, mounted at the end of a horizontal mandrel, was rotated within the casting cutting away the metal and advancing slowly as it did so. Such a simple machine could be driven by means of a water-wheel, a tread-mill, a windlass or a horse gin. But the cannon was very far from being a precision weapon and the boring machine remained therefore a crude and inaccurate tool: why bother, one might ask, to strive for accuracy of bore when the difficulties of cannon casting and the uncertainties of gunpowder manufacture imposed such narrow limits on the range and accuracy of cannon?

Almost certainly it was a multiplicity of factors that caused the sudden development of machine tool technology in the period 1790–1825; one such factor was plainly the steam engine. Watt had demanded wholly new standards of accuracy in the construction of his efficient condensing engines. The need to have a working cylinder that was a true right cylinder and of uniformly circular cross section led to the famous (and chequered) collaboration between James Watt and John Wilkinson, the ironmaster who had built a large boring machine of unprecedented accuracy. On the other hand the absence in 1784 of a suitable planing machine had compelled Watt to exercise his genius and invent the parallel motion. It seems certain therefore that the rapid improvement of the steam engine and its increasingly wide application must have resulted in growing pressures and incentives to develop the machine-tool industry. Only with industrial machine tools could large and really efficient steam engines be built.

We might therefore reasonably expect to find that a machine tool industry grew up wherever steam engines were made. But in one very important instance this did not happen. From about 1800 to about 1870 the most efficient steam engines in the world were manufactured in Cornwall but no machine tool industry developed in that county. The centres of the machine tool industry during the nineteenth century were (in England) London, Birmingham, Manchester and Leeds and (in Scotland) Glasgow. In other words machine tool making seems to have flourished where there was a diversity of industry and a wide variety of skills available. Birmingham was—and still is—a city with many trades, but as regards the design and manufacture of machine tools it seems likely that its traditional association with fire-arms was possibly the most important single factor. In the cases of Manchester, Leeds and perhaps Glasgow we might not be too far wrong in relating

the machine tool industries to the development of the manufacture of standardised textile machinery.

With this very varied background, comprising in different
degrees the whole range of industrial practice in England,
the rise of the machine tool industry still depended on certain
dominant personalities, the analogues of and in some ways
curiously similar to the Arkwrights, the Strutts and the Cromp-
tons of the textile industry. Outstanding among these early
fathers of the industry were Joseph Bramah and Henry Maudslay. Bramah, the son of a Yorkshire farmer, was a prolific and
able empirical inventor. Among his better-known inventions
were the lock that bore his name, the familiar beer engine
and the hydraulic press that foreshadowed the development of
hydraulic power transmission techniques in the nineteenth
and twentieth centuries. Like other engineers of the time, he
dabbled unsuccessfully with the direct rotative steam engine.
But Bramah was not only an inventor, he was a manufacturer
and the problem of producing large numbers of identical
locks led, inevitably, to the use of specialised machinery. In
devising the machinery Bramah had the assistance of a young
man, his works foreman, whose abilities equalled and perhaps
exceeded his own.

This young man was Henry Maudslay, and it was he who
made the first industrial lathes. Maudslay realised, very
clearly, that if they are to cut metal accurately, industrial
lathes must be absolutely rigid and therefore made wholly of
iron or steel, that the spindles and dead-centres must be
precisely aligned and that the surfaces over which the slide
rests and cross-feeds move must be true, perfect planes. Only
under these conditions would all the motions be precisely and
geometrically determined and accurate cutting be possible on
the scale required to produce the major components of well-
engineered steam-engines. Maudslay, who had left Bramah
and set up on his own account in 1797, insisted that each of his
workmen engaged in the manufacture of lathes should have
a perfectly flat standard plane beside him on his bench and
should use it constantly to maintain the highest precision
standards. All these principles were manifest in the screw-
cutting industrial lathe, fitted with an accurate lead screw and
slide rest with cross feed, that Maudslay produced in 1797.

E

The introduction of industrial machine tools is related to what is now known as mass production. In a sense mass production is an old technique for which the requirements are only a uniform standardised product, a large market, and the division of labour. The manufacture of pins, which so impressed Adam Smith, was clearly an early if very simple instance of mass production. During the eighteenth century, however, the manufacture of clocks, locks and other rather more complex commodities led to the development of special machinery of which that devised by Bramah and Maudslay was a good example. In 1798 the American engineer Eli Whitney commenced the mass manufacture of muskets machined to such a high standard of accuracy that the parts were interchangeable. But perhaps the most dramatic instance was the famous Portsmouth block-making machinery for the mass production of pulley blocks.

During the Napoleonic wars one of the major problems in the maintenance and refitting of the blockading British fleets was the supply of pulley blocks. An immense number of these components, common to all ships, was required. Accordingly Marc Isambard Brunel, father of the famous railway engineer, designed a set of machines for the mass production of pulley blocks. He was assisted in this work by General Samuel Bentham and Henry Maudslay. Rectangular blocks of wood were cut out, drilled, slotted and shaped in sequence and the pulley block then completed, each operation being carried out by a specially designed machine. The resulting economy of manpower was such that ten men could do what had previously required over one hundred. These famous machines can be seen at the Science Museum in London.

With Maudslay, the industrial machine tool industry, properly speaking, began. It was carried on by Richard Roberts, the brilliant Welsh engineer who, in 1817, was one of the first to build a planing machine; by Joseph Clement and James Nasmyth all of whom had worked for Henry Maudslay, and by James Fox of Derby who had not been taught by Maudslay. But the perspectives of history should not lead us to believe that the nature of the new industry was immediately apparent to contemporaries, It was not. A perusal of the dictionaries and encyclopædias of the time will show that the phrase, and therefore the concept of, 'machine tool' was not known. In for example Abraham Rees' magnificent *Cyclopædia*, (1819), there is, under the article 'lathe', an account of Mr Maudslay's screw cutting lathe; under 'cannon', a description of a boring machine and, under 'machinery', a long and comprehensive account of the Portsmouth block-making machines. But the only other machine tool that would be accepted as such to-day was a tube drawing machine, described under 'pipes'.

The next significant development in the revolutionary period 1790–1825 was a single individual invention that was to have far reaching consequences in technologies undreamt of at that time. This was the loom invented by J. M. Jacquard in 1801. It was not without precedent; in fact it represented the final resolution of an old problem: that of weaving, economically and quickly, a fabric with a pattern that repeats itself.

Broadly speaking there were two ways of patterning woven fabrics; the cheaper way, suitable for cottons, for example, was by printing. The technique of printing was soon brought to a high degree of perfection in the wake of the great textile inventions of the late eighteenth century and the advance of chemical science that took place at the same time. The other way, essentially the same as embroidery, was to use different coloured warp and weft threads and to vary the order in which these passed under and over one another. This was the technique used in the case of silk fabrics and to produce the patterns a draw-loom was used. By raising certain warp threads in strict succession and then returning to exactly the same sequence and repeating this procedure indefinitely a repetitive pattern can be woven of the sort found on neck-ties, for example. In practice the raising of the selected warp threads was carried out by a draw-boy who had the rather thankless and certainly unskilled task of raising the sequences time after time as the pattern was woven. It was the essence of Jacquard's invention that he wholly mechanised this procedure thereby not only saving in labour and time but also, as it happened, opening the way to new technologies.

The principle of the Jacquard loom is very easy to grasp (Figure 29). The warp threads to be raised during a traverse of the fly shuttle from one side to the other are determined by holes punched in a rectangular piece of stout cardboard.

This piece of board is held vertically and is 'read' or 'sensed' by being probed by an array of thin horizontal steel rods. These rods are mounted with springs at the other end and if one of them encounters a hole it passes through it so that the

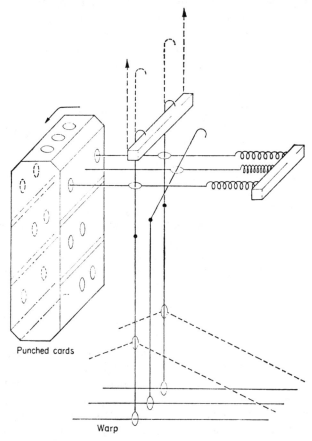

Punched cards

Warp

Figure 29.   *The Jacquard loom.*

spring remains uncompressed; if there is no hole the rod cannot pass and the spring at the other end is compressed. At a certain distance along each rod there is a loop through which a vertical wire passes. These vertical wires have hooks at their top ends and support the warp threads at their lower ends. If a horizontal rod passes through a hole the vertical wire is displaced (to the upright position in the diagram); if there is no hole the wire is not displaced. A horizontal bar is arranged so that only the hooks on displaced wires engage it in such a way that when the bar is raised the displaced wires, together with the warp threads they support, are raised too. The undisplaced wires and their warp threads are not raised. After the appropriate sequence of warp threads has been raised in this way and the fly shuttle carrying the weft has made its traverse, the next punched card is substituted, a fresh sequence of warp threads is raised and the fly shuttle returned. This procedure is repeated indefinitely. As the pattern is repetitive the cards bearing the sequence of holes are connected together to form an endless belt that is fed through the 'sensing' or 'reading' mechanism of the loom.

The Jacquard loom was immensely and immediately successful* and, in basic principle, it is in use at the present day. More, the idea of 'programming' an automatic machine in Jacquard fashion proved very fruitful. The presence or absence of a hole corresponds to 'on' or 'off', or in binary notation, to 0 or 1. This way of coding information and then 'reading' it by means of probes was the starting point of modern computer technology, for it directly inspired Charles Babbage, the founding father of the computer. As early as 1847 Richard Roberts had designed a multiple spindle drilling machine, controlled by a Jacquard mechanism, for drilling the many girders required for the Britannia tubular bridge then being built across the Menai straits.

The last significant development of this period that we shall discuss, albeit briefly, was the rapid advance in electrical science. In 1796 Alessandro Volta discovered that if two pieces of dissimilar metals are brought into contact and then separated they acquire opposite electrical charges. He then showed how, by arranging pairs of bimetallic discs so that each pair was separated from its neighbours by a conducting

* Some seven years ago the author obtained from an old mill in Maccles-field a Jacquard loom that had been in use six months earlier. This loom is virtually indistinguishable from the one depicted in A. Barlow's *History and Principles of Weaving* (London 1878), p. 140, and reproduced in A. P. Usher's *History of Mechanical Inventions* (Harvard University Press 1962), p. 296. To judge by the personal details of dress and hair style the original engraving must have been made about 1845.

liquid, a continuous flow, or current, of electricity could be obtained. He had, in fact, invented the 'Voltaic pile', or electric battery; and it initiated a dramatic period of discovery in electrical science. In 1819 H. C. Oersted, a Danish physicist, placed a horizontal wire, running along a north-south line, just above and parallel to a magnetic compass needle. When an electric current flowed along the wire Oersted noticed that the needle turned either towards the west or the east depending on the direction of the current. This discovery constituted, in effect, an efficient detector of electric currents: a 'galvanometer', as it was soon called. Also it made the electric telegraph and, as we shall see, much else, possible. A Voltaic pile, a switch or key, a double length of wire and a galvanometer at the far end were sufficient for the communication of suitably coded messages between the keying point and the distant galvanometer. There had been attempts to use electricity for signalling before, but they were unsuccessful for static electricity and Leyden jars were not really satisfactory. Now at last there was an effective means; and there was a strong incentive, too, for the developing railways needed some systems whereby messages could be sent along the track to warn stations of the movement of trains.

This was another invention that Bacon would have approved. But he would have been less happy about one of the very earliest applications of Voltaic electricity, which was to explode land mines. A charge of gunpowder could be detonated at a distance by an electric current causing a piece of thin wire, buried in the powder, to become red-hot. No great genius was required for this invention, but it did have one interesting feature. If the device was to work in wet weather, or on damp ground, the wires had to be insulated to prevent short-circuits. They had, in other words, to be coated with an insulating substance. So wires coated with gutta-percha were used; and in this way, early in the nineteenth century, the very familiar electric flex was born.

FRENCHMEN AND ENGLISHMEN: TECHNOLOGY AND INDUSTRY

The majority of the names we have mentioned so far in this chapter have been those of Englishmen and Scotsmen. This was not the result of narrow chauvinism: far from it. It is simply because at this time the most immediately applicable and therefore profitable technology was being done in the United Kingdom. The facts cannot be disputed; Britain was carrying out an industrial revolution just as France was carrying out a political, social and military one. But this does not mean that technology was not being forwarded on the Continent and in France in particular. It was indeed, although its application was less immediate and of more academic interest than that being done in Britain. It is a curious but indisputable fact that more emphasis was put on the advancement of technology as public policy in France than in Britain.

During the years 1790–1825 France had more scientists and technologists of the first rank than any other nation had ever had over a comparable period of time; on a *per capita* basis the French record has not been equalled even up to the present day.* There were, of course, several reasons, apart from the genetic mysteries of genius, for this. The intellectual supremacy of France in the eighteenth century was matched by her military and political domination of Europe. Germany, Italy and Spain were in eclipse while the rising power—Britain— was engaged in overseas adventures: exploration and colonisation. French technological and scientific genius received a new and powerful impetus when the revolution, born of eighteenth-century rationalism, broke out. A new social order, with the ideal of equality prominent, meant better opportunities for talented young men; educational institutions like the École Polytechnique (1794) gave first class training in engineering and physical sciences; high national morale coupled perhaps with the example of the dashing new military tactics of the Napoleonic generals brought about a new professional style in science and technology.

The excellence of French technology can easily be confirmed by anyone who takes the trouble to look at such works as

* Among the famous names we may instance: Lavoisier, Laplace, Prony, Monge, Chaptal, Montgolfier, Clément, Desormes, Dulong, Petit, Leblanc, Jacquard, Burdin, Poncelet, Hachette, Christian, Gay-Lussac, Biot, Haüy, Fresnel, Guyton de Morveau, Fourcroy, Berthollet, Proust, Ampère, Savart, Arago, Sadi Carnot, Fourier, Dupin, Coriolis, Cauchy, Cuvier, Lamarck.

Riche de Prony's *Nouvelle architecture hydraulique* (1790, 1796) which was admitted even in England to give the best contemporary account of the steam engine; Lanz and Bétancourt's *Essai sur la composition des machines* (1808), which set out a system of ten classes of mechanisms that could relate the different forms of circular and reciprocating motion, thereby inaugurating the science of kinematics: J. N. P. Hachette followed this classification in his *Traité élémentaire des machines* (1811). Other notable works of this decade were A. Guenyveau's *Essai sur la science des machines* (1810), C. L. M. Navier's magnificent revised edition of Belidor's *Architecture hydraulique* (1819) which was virtually an original work in its own right and A. M. Héron de Villefosse's *De la richesse minerale* (1819) which was a comprehensive survey of mining technology in Europe and a worthy successor to Gabriel Jars' work. In the following decade J. A. Borgnis's *Théorie de la mécanique usuelle* (1821) and G. J. Christian's *Traité de la mécanique industrielle* (1822–1825) appeared.

The modern reader who cares to contrast these important, indeed seminal, French works with such English contemporaries as William Emerson's *Principles of Mechanics* (1770), John Banks' *On Mills and Mill Work* (1790, 1792), Olinthus Gregory's *A Treatise on Mechanics* (third edition 1815) and Thomas Tredgold's *Tracts on Hydraulics*, will be forced to conclude that the English works showed little or no advance on Desaguliers and as regards scientific and mathematical ideas no advance at all on Newton's *Principia mathematica* (1687). Quite literally the English textbooks belonged to an earlier century.

The excellence of French technology, however, was not merely a matter of textbooks, however advanced. Gaspard Monge had established projective geometry and founded the art of engineering drawing. It was from the work of Monge and N. L. M. Carnot that Lanz, Bétancourt and Hachette developed their pioneer classification of kinematics. We have already discussed J. M. Jacquard's contribution to textile technology and we have briefly mentioned French leadership in ballooning and parachuting. In power technology a number of interesting versions of the air-engine were invented, including the 'pyreolophorus', a novel type of buoyancy engine due to Cagnard Latour and an engine in which the mixture of air and gas was ignited in the cylinder by means of an electric spark (1801). This last was due to Phillipe Lebon who, like Cayley and the Niepce cousins, used a pump to charge the cylinder with (relatively) compressed air and gas.

Even in the field in which the British have always prided themselves, namely shipbuilding, the French were pre-eminent. The late G. S. Laird Clowes, writing of British shipbuilding during the Napoleonic wars, remarked that much was learned from captured prizes for the French had long treated ship-building much more scientifically than had the British. In the construction of buildings and in making roads and bridges, canals and harbours, there is no reason to doubt that the French engineers were as good as, if not well ahead of, all others. August Coulomb had made fundamental contributions to the theory of structures, while the famous Ecole des Ponts et Chausées had been founded in the middle of the eighteenth century. It was during this period that French engineers and scientists established the theoretical bases of civil engineering.

Finally we note that no country in the world had ever given science and technology so important a place in public policy as had revolutionary France. Prizes were instituted, the Ecole Polytechnique recruited its students by competitive examination from among the most able boys in France. Poor students were supported by the State and the professors were the leading engineers and scientists of France. Technologists and scientists were honoured and new inventions were backed with public funds. The contrast with England could hardly have been sharper. The Royal Society at that time was a gentlemen's club, the two universities were almost moribund, open only to a select few and bound by archaic statutes. There was nothing to rival the Ecole Polytechnique and the advancement of science and technology was a matter of private interest and amateur effort.

It is no more than a truism to say that if State encourage-ment of technology and science coupled with first-class educat-ional facilities and a galaxy of technological and scientific talent could have done the trick, then the industrial revolution must have taken place in France rather than in England; or, at the very least, France should soon have overhauled England in the early years of the nineteenth century. That she did not is a clear enough indication that—modern pundits notwith-standing—scientific and technological factors are not, in themselves, sufficient, or even perhaps the most important, for rapid industrial progress.

In trying to find reasons why France did not overtake Britain we must be acutely aware that all explanations we put forward are open to the accusation of hindsight. This does not, however, exempt us from the responsibility of trying to account for this paradox.

In the first place although France had a respectable coal-mining industry, second only to that of Britain, she had nothing to compare with the great non-ferrous metal mining industry of Cornwall. Mining, especially of the latter sort, had long been associated with progressive technology. This was particularly true of eighteenth-century England where mining was linked to a close network of related and complementary technologies: metallurgy, refining, iron founding, power and transport. These technologies stimulated others in turn so that the great system inter-locked and ensured general progress.

Secondly, it is possible that the very virtues of the French system militated against economic success. France was, as Britain was not, a centralised country. Paris, the capital, was also the only important centre. It is natural for functionaries to favour centralism: it is so much tidier to have everything run from one big central city. Accordingly talent was drawn to Paris and the Ecole Polytechnique. Once in the capital the able young man was given every incentive to stay there. In England on the other hand there were several important centres besides London. In the midlands, around Birmingham, a notable group of technologists and scientists came together to form the Lunar Society, while in Manchester the Literary and Philosophical Society formed the centre for such men as John Dalton and William Henry; and later for Peter Ewart, William Fairbairn and J. P. Joule. In Scotland there were vigorous intellectual communities in Edinburgh and Glasgow, while in Ireland Dublin could boast of such men as William Higgins, the chemist, and later William Rowan Hamilton, the mathematician. All these active centres meant a wider and more effective diffusion of talent than was possible in France.

Thirdly, it may not be entirely subjective to discern a broad difference between the practice and personnel of technology in England and in France at this time. English technology seems to have been generally more empirical than French which was more theoretical, more scientific. Did this perhaps reflect not so much a difference in national temperament and in scientific organisation as the existence of a class in England that had few or no representatives in France? A class, that is, of technicians broadly self-educated and self-made: Arkwright and Jedediah Strutt, Trevithick and Thomas Hewes? Men who were in very close touch with immediate and practical requirements and had both the ability and the means to effect fruitful inventions. In France, on the other hand, the engineer appears more typically as a mathematician, a Monge, a Carnot, or a Prony. And while such men were well able to discern the future trends and possibilities inherent in technology and science it by no means followed that they were able to spot the immediate winners—and a progressive economy depends as much on short term as on long term success. Certainly it seems to have been the case that the British technologists could innovate effectively and economically while French ideas, although brilliant, were often applicable only in the distant future.

But this is as far as we can go in debating this fascinating and important question. Other, non-technological, factors were no doubt equally relevant—the favourable geographical, political and geological circumstances of England, her refusal in effect to divert too great a portion of her technological resources to economically sterile military matters—but these we must leave to others to discuss and evaluate. The final answers are not yet written.

STEAM-ENGINES AND COSMOLOGY

We have already pointed out that the ending of Watt's stranglehold on the steam-engine business led to the rapid development of the mobile high-pressure engine through the genius of Richard Trevithick and his contemporary Oliver Evans and, in a more orthodox tradition, a progressive improvement in stationary pumping engines. In the case of the pumping engines progress was not immediately apparent for it seems that the withdrawal of Watt's influence in Cornwall had an unfortunate effect on standards. Two things however were to change this and set the Cornish engine-builders firmly on an upwards path of improvement and increased efficiency.

It began with Arthur Woolf's compound engine. Woolf, another self-made engineer, patented his principle in 1804. The experimental data on which it was based was defective but in broad outline the idea was sound. It amounted to a Hornblower type compound engine using Watt's separate condenser and using the double acting principle. When the first of these engines was set up at Wheal Abraham mine in Cornwall the results were a startling improvement on what had gone before. A duty of some 56 million ft. lb. per bushel of coal was achieved: roughly twice as good as the best that Watt's low-pressure engines could achieve.

Woolf's engine was important not only because of its economy but because its performance was compared with other engines as part of a systematic policy of recording the performances of all engines in Cornwall. In an effort to reverse the immediate post-Watt deterioration in performance the Cornish engineers and mine owners resolved to publish the details of the performances of their engines so that factors leading to efficiency could be identified and developed while inefficiencies could be eliminated. Accordingly from 1811 onwards Joel Lean, a Cornish engineer, collected all the relevant details from the vast majority of engines working in that county—which meant in effect from all the best engines in the world—and published them in his *Engine Reporter*. These results led to the recognition of the superiority of Woolf's engine and, even more important, of the high-pressure, expansively operated engine, of which type Woolf's engine was an example. Month in month out, year in year out these results were published and they helped not only to establish the truth of the generalisation that high-pressure expansively operated condensing engines were most efficient, but also to stimulate Cornish engineers to improve performance by all means at their disposal: better designs of valves, of thermal lagging, of furnace design and so forth until this highly competitive 'open society' of engineers achieved some astonishing engine performances: 87 million ft. lb. per bushel in 1828 and then finally, a reported 125 million in 1835. There had been nothing like this before in the whole history of power.

Neither the Lean reports nor the performance of the Woolf engine in particular were ignored in France or in England. When the wars ended in 1815 Woolf's old partner Humphry Edwards went to France and started to build Woolf type engines in that country. Edwards was immediately and consistently successful and it soon became apparent in France, as in England, that the high-pressure expansively operated condensing engine was more economical as well as superior in power/ weight ratio. Before very long the Woolf type of compound was dropped: it was rather too complex and comparable performances could be obtained from single cylinder, expansively operated high-pressure engines. But the point had been made: high-pressure operation was 'in'.

It was during this post-war phase, when France was assessing the amazing achievements of English technology and the steam-engine in particular that a young military engineer, a son of the illustrious N. L. M. Carnot, took time off to survey the whole field of heat and the power that it can develop if properly applied. In 1824 the young Sadi Carnot published his observations in a quite remarkable short book called *Reflections on the Motive Power of Fire*. It is no exaggeration to sum this book up as the most *original* work of genius in the whole history of the physical sciences and technology.

He begins by remarking that recent scientific advances have confirmed that heat is the great motive agent, the great source of power, of the universe.

> It is to heat that we must attribute the great and striking movements on the earth. It causes atmospheric turbulence, the rise of clouds, rain and other forms of precipitation, the great oceanic currents that traverse the surface of the globe and of which man has been able to harness only a tiny fraction for his own use; lastly, it causes earthquakes and volcanic eruptions.
>
> From an immense natural reservoir we can draw the motive power we need; nature in offering us all sorts of combustibles has given us the means of generating at any time and anywhere the latent motive power. To develop that power, to appropriate it to our own use is the purpose of fire-engines.

This is extremely clear if a trifle exaggerated. The fire-engine has given us a new and profound insight into the nature and working of our universe. Technology has, once again, changed man's consciousness of his world, radically and irrevocably. If, to seventeenth-century philosophers the universe seemed like a gigantic piece of clockwork, to nineteenth century thinkers it was to appear to have many of the attributes of a heat-engine. On a practical plane, the steam-engine has transformed the conditions of life and the civil economy of England:

> To take away England's steam engines to-day would amount to robbing her of her iron and coal, to drying up her sources of wealth, to ruining her means of prosperity and destroying her great power. The destruction of her shipping, commonly regarded as her source of strength, would perhaps be less disastrous for her.

These were astonishingly perceptive observations for 1824. But Carnot was not concerned so much with the economic, or even the cosmological implications of heat. He was interested in heat-engines in general and it was here that his perceptions rose to the highest pitch of genius. Although there is, he points out, a complete theory of water power, applicable to all water-engines, there is no general theory to account for heat-engines which in recent years have shown such a dramatic increase in efficiency. Such a theory is urgently needed and he proposes to put one forward.

Exactly as in the case of water power wherever there is a difference in level—or temperature—there is the possibility of generating motive power. In other words where you have a hot body and a cold body you can derive useful power by means of a heat-engine.

We must recall that at this time the analogies between water and heat power had become very close. Although there had been little or no similarity between one of Parent's water-wheels and a Newcomen engine, by 1824 engineers were trying to devise direct rotative engines that could be driven by either steam or water, or even reversed and used as water pumps; we had reaction engines driven by water (Barker's mill) or by steam (von Kempelen's engine) and we had the versatile column-of-water engine which was really a high-pressure 'steam' engine driven by water. The incentives to think about heat-engines in the same terms that one thought about water-engines were therefore quite strong. And they were considerably strengthened by the commonly accepted doctrine of the day, that heat was of the nature of a very subtle, or fine fluid called 'caloric'. This fluid will tend to flow from a hot body to a cold body just as water will flow from a high level, or pressure, to a low level, or pressure; the analogue of pressure or head in the case of heat being the temperature.* It was therefore very easy for Carnot to envisage heat-engines working in much the same way that water-engines worked: by means of a fluid flowing from a high level, or hot body through the engine to a low level, or cold body (condenser). Dr A. J. Pacey has pointed out that the presence of the condenser, as a cold body, made it very easy for Carnot to envisage the heat-engine working in this fashion. The Trevithick engine, without a condenser, would have been less suggestive to the imagination in this respect.

With these ideas in mind Carnot sketched in very briefly but convincingly the conditions under which the maximum power could be generated from the transference, or 'fall' of heat from a hot to a cold body; from, that is, a boiler to a condenser. In the case of water-engines the prime conditions for maximum efficiency were that the water should enter the machine without turbulent impact and should leave it without velocity. If the machine is mechanically perfect and there are no losses in its working then its efficiency will be 1 and it will deliver enough power to enable a perfect pump to restore all the water to the driving source. This argument had been formally used by Déparcieux in 1752 but it was implicit in all calculations of the efficiency of water-engines. If, like Carnot, we now extend

* A number of scientists and engineers refused to commit themselves to the doctrine of caloric, conceived as a subtle fluid. Since this substance could neither be seen, nor weighed nor detected by any other means than its overt effects they preferred to reserve judgment on its actual existence. This did not affect their arguments however for they, like the calorists, accepted the fundamental axiom that heat is always conserved. This meant that in driving a heat-engine heat is conserved and not destroyed or transformed, just as water is conserved in driving a water-wheel: as much water comes out of the wheel as went over the top.

the argument to the case of heat-engines we see at once that a perfect heat-engine must be one that delivers enough power to pump back all the heat to the hot body; or in other words, to restore the status quo.

The prime conditions for the maximum efficiency of a heat-engine are that the working substance—air, steam or any other vapour—has the same temperature as the hot body when the two are in contact at the beginning of the expansion and the same temperature as the cold body or condenser at the end. If this is not the case, if there are temperature differences, then the possibility of generating a certain amount of motive power must have been missed. There must have been a useless flow of heat: the thermal equivalent of a wasted flow of water, or of turbulence.

The working substance in the cylinder is first of all expanded at the temperature of, and in contact with the hot body. Then, after a short distance it is isolated from the hot body and the engine works 'expansively'. As the volume of the isolated gas or vapour increases so its pressure and its temperature fall. This process is known as 'adiabatic' expansion, and it continues (Figure 30) until the temperature has fallen to that of the cold body or condenser. The piston has now reached the end of the cylinder and in order to repeat the cycle it is necessary to return it to the other end and put behind it the required volume of hot air or steam at the starting pressure. This can be done in two ways: either by heating up a fresh quantity of air, or of water and then evaporating it, or by compressing the expanded air or steam in the cylinder, keeping it in contact with the cold body all the time and in effect squeezing the heat out of the working substance into the cold body. Then having compressed it sufficiently at the low temperature, the working substance must be isolated from the cold body and compressed 'adiabatically' until its temperature and pressure are back at the starting point. This second method is more efficient since it avoids bringing hot and cold bodies together. For we see that, following Carnot's initial insight, a straightforward flow of heat is to be avoided as it must entail some loss of motive power.

The Carnot cycle, so defined, represents the most efficient possible operation of a heat-engine. If the prime conditions are adhered to rigorously and if there is no useless flow of heat the engine can be run in reverse so that the net effect will be a consumption of motive power and the restoration of all the heat to the hot body; in effect all the heat will be pumped back from the cold body to the hot one. A perfect column-of-water engine in which the load increases to such a point and in such a way that it drives the engine as a pump will give some idea of the Carnot argument at this stage.

The proof that an engine working on a Carnot cycle is the most efficient engine possible is quite simple. If we imagine, for the sake of argument, that there is a still more efficient engine

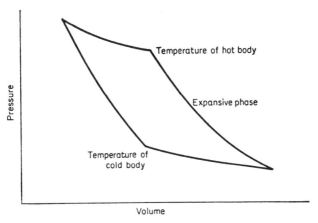

Figure 30.

then we have only to imagine this machine used to drive a Carnot engine in reverse to see that the net effect will be the restoration of more heat to the hot body than is extracted from it, or let down, to drive the more-than-perfect engine. But in this case, since heat is always conserved, the consequence must be that heat will accumulate in the hot body and this implies that perpetual motion can be obtained. An idea that, as Carnot remarked, is repugnant to science and to common sense.

For a given difference in temperature between a hot body and a cold one therefore, no engine can possibly be more efficient than one that is completely reversible. It follows that all such engines must have exactly the same efficiencies whether the working substance is air, steam, or any other gas, vapour, liquid or solid. This demonstration, so clear, so unambiguous in theory, enabled Carnot to draw very important conclusions for the physics of gases; these do not concern us here for they belong to the history of science rather than that of technology. We should, however, note the general scientific implication of Carnot's ideas. Heat is associated with all the physical and chemical changes in nature; at one end of the spectrum, as it were, we have the simplest imaginable case of pure conduction of heat from a hot body to a cold one without any other kind of change; at the other end of the spectrum we have the equally idealised case of the Carnot engine whereby the flow of heat from a hot to a cold body results in the appearance of the maximum amount of mechanical energy. All the other cases are contained between these two extremes. Carnot's ideas, derived directly from power technology, had therefore given rise to a wholly new way of looking at the phenomena of heat.

The general experience of engineers, certainly since 1815, had been that the most efficient engines were high-pressure, expansively operated condensing engines. Some engineers, but by no means all, had gone further and claimed that high pressure *per se* resulted in higher efficiency. There were, in any case, three perfectly good reasons for the superior performance of the most recent engines. Firstly, the expansive principle becomes more effective as the pressure of the steam is raised; secondly, high-pressure steam-engines being smaller than low-pressure ones of the same power tend to lose less heat by radiation and convection, and must therefore be more economical; thirdly, high-pressure engines tended inevitably to be more modern than low-pressure ones and therefore incorporated numerous detailed improvements that enhanced their performance. An engineer building a new, high-pressure engine would incorporate new and superior valve mechanisms, better thermal lagging, a better boiler and furnace and so on. The steady improvement in engine performance, revealed by the Lean reports, confirmed that engineering standards were rising and with them the efficiencies of engines.

* The adiabatic cooling of air when it expands is the cause of several important meteorological phenomena. The converse effect, the adiabatic heating of compressed air is familiar to all those who have operated a bicycle pump.

Carnot was deeply impressed by the superiority of the high-pressure steam-engine, which he thought must be due to some simple basic principle. The temper of his mind was comprehensive and synthetic, rather than specialised and analytic and the analogy of water power suggested to him that the key to the superior performance of high-pressure engines lay in the extent of the fall of the moving agent. The most efficient hydraulic engine is plainly the one that harnesses every inch of fall of water from the highest available point to the lowest. The high-efficiency column-of-water engine exemplifies this principle very clearly. Applying the same argument to steam- or heat-engines it would follow that the best engine would be one that harnessed the greatest fall of heat, or caloric. Coal burns at about 1,000°C and the temperature of the cold water in the condenser will be about 15°C so that the total fall is about 985°. But of this, the total 'fall' available, the Watt-type engine makes use of about 85° only while an engine working with steam at six atmospheres and with steam at a temperature therefore of about 160°, would use 145°. The high-pressure engine in other words uses nearly twice as much of the available fall as the low-pressure one. So, he remarks:

> It is easy to see now what are the reasons for the superiority of the so-called high-pressure engines over the low-pressure variety; *the advantage lies essentially in the ability to harness a bigger drop of caloric.* Steam generated at a higher pressure is also at a higher temperature, but as the temperature of the condenser is always more or less the same, the fall of caloric is evidently greater.

Was Carnot justified in making this rather sweeping pronouncement? In terms of the available knowledge, theoretical and practical, of his time he was not. The three major practical factors we have mentioned were quite sufficient to explain the superiority of the high-pressure engine. We can see, with the advantages of subsequent knowledge, that the basic 'thermodynamic' reason that Carnot had intuited was, over the relatively small temperature range involved, too small to be significant. Or, in other words, the improvement in the performance of high-pressure steam-engines between 1800, or 1815 as far as France was concerned, and 1824 was due mainly to improvements in design, to the increased use of the expansive principle and to a reduction in size, and only incidentally to the 'thermodynamic' factor that these engines worked over a greater temperature range. The latter factor then was masked by the overall rise in performance that was much greater than thermodynamic theory would predict.

In terms of intuition, however, Carnot was abundantly justified: he turned out to be right. And he went on with brilliant clarity to expose an unnoticed drawback to steam as a motive agent. We have commented on this particular point before (page 108) but as it is so important historically and as no writer has, as far as I know, previously called attention to it, I will repeat Carnot's argument briefly:

The enormous expansion of water, some 1,800 fold, when it is converted into steam by boiling at normal atmospheric pressure was the main reason why it was so eminently suitable as a 'working substance' for heat-engines. In other words, a great expansion was easily obtainable for a moderate temperature rise. This could not be rivalled by, for example, air which expands by barely one third over the same increase in temperature (say 15° to 100°C). Correlatively if the steam was prevented from expanding against the pressure of the atmosphere (15 lb. per square inch) by being confined in a very strong vessel, the increase in pressure would be enormous compared with the rise in temperature of the steam.

The essence of Carnot's superb insight was this: he saw that what had, up to his day, been the great advantage of steam— its property of enormous volumetric expansion or of tremendous increase in pressure for a moderate rise in temperature, the very property that had made Newcomen, Watt, Trevithick and Woolf engines possible—must in the future prove a disadvantage. The very fact that the pressure increased much more rapidly than the temperature meant inevitably that it would be much more difficult to harness the total fall of temperature from about 1,000°C of burning coal to the temperature of cold water, or of the atmosphere:

> one of the most serious disadvantages of steam is that it is not possible to employ it at high temperature without using containers of extraordinary strength. It is not the same with air, for which there is no fixed relationship between the pressure and the temperature. Air thus seems to be better than steam for

harnessing the motive power of falls of caloric from high temperatures; perhaps in the lower range steam is more convenient. Conceivably, the same quantity of heat could be made to act first on air and then on steam. The air should still have, after its use, a high temperature and instead of exhausting it straight out into the atmosphere it should be circulated round a steam boiler as if it came from a furnace (firebox).

The use of atmospheric air for generating the motive power of heat raises very serious difficulties in practice; but these are not perhaps insurmountable. If they could be overcome air would be without doubt found to be much more effective than steam.

This was surely the most penetrating and stimulating insight in the whole history of technology. It is difficult to find a rival of even approximately the same quality. One possible candidate might be Matthew Boulton's prophetic remark, made in 1769 to James Watt: '. . . . it is not worth my while to manufacture your engine for only three counties; it is well worth my while to make it for the whole world.' Boulton, at any rate, foresaw the universal application of the heat-engine long before the actual event. But even so, Boulton's foresight, or prevision, does not rival that of Carnot. For Carnot's conclusion was based on strictly scientific principles; it was not a piece of inspired guesswork, a flight of fancy or an example of superb business acumen. It led directly to the high-efficiency air-engines developed towards the end of the century, for Carnot's theory supplied the fundamental principles that were absolutely necessary to make the air-engines dreamed of by men like Cayley really practicable propositions. For the present it is enough to note that the history of the heat-engine falls into two quite distinct periods: before Sadi Carnot, when the emphasis was on obtaining the maximum expansion at a reasonably low temperature, and after Carnot when it was confirmed that the maximum efficiency entailed harnessing as much as possible of the total temperature range from that of the burning fuel to that of the condenser, or the surrounding air.

Carnot died, tragically, of cholera in 1832 and his work, which amounted to the establishment of a new and fundamental science (thermodynamics) and a recasting of the technology

of the heat-engine, was almost completely ignored for ten years
or so. There were unusually good reasons why the best engineers, mathematicians and physicists should have overlooked
it.

In the year that Carnot published his masterpiece, 1824,
Burdin gave an account of an invention that he called the high
speed 'turbine' This was a water engine so designed that the
loss of *vis-viva*, or rather of energy, due to the turbulent impact
of water on blade was reduced to a minimum and at the same
time that the water should leave the machine without appreciable velocity. Basically the engine consisted of two concentric
cylinders with the annular space between them occupied by
doubly curved blades fixed to the inner, moveable cylinder.
Water enters the machine in a direction parallel to the common
axis and causes the inner cylinder to rotate by acting smoothly
on the doubly curved blades. It leaves the machine at right
angles to the common axis, having been diverted by the curved
blades, and in the opposite direction to that of rotation of the
inner cylinder. In this way the highest possible efficiency is
obtained and the machine works at high speed. It does not
suffer from the disadvantage of the overshot or breast wheel
that it must run slowly in order to be efficient.

Burdin's turbine was, in fact, a highly advanced machine,
designed in accordance with the theory expounded by Lazare
Carnot for the optimum efficiency of water-engines. It is,
therefore, strictly speaking unhistorical to describe, as is often
done, early water machines with curved blades as 'turbines':
there were no turbines before 1824. In the event the requirements that Burdin postulated were beyond the engineering
industries of his time and it was not until the middle of the
nineteenth century that the first practicable water turbines were
built. They were to supersede the column-of-water engine as
the best means of developing all the power inherent in a really
big fall of water. And although they were often used to develop
direct rotative power—in 1905 the big water-wheel at Quarry
Bank Mill, Styal was replaced by a turbine that is still there
to-day—it was not until the development of electric power
that the water turbine really became important. And it was
with the rise of the electrical industry that the offspring of
Burdin's turbine, the steam turbine of Sir Charles Parsons,

also became important. Later still the gas turbine, or jet engine appeared.

With Burdin's remarkable invention we leave this revolutionary epoch, 1790–1825, when so many of the technologies of the present day were first established, clearly and automatically.

Chapter 5

The High Tide of Progress

Even before the end of the great wars that so effectively cut industrialised Britain off from the continent of Europe, there were signs of growing dissatisfaction with the state of science in England. The universities, devoted to a pedantic conning of Newton's *Principia*, had long since fallen behind the great Continental schools that had developed the powerful Leibnizian form of calculus, as opposed to Newton's fluxions, and established a generalised science of mechanics—*mécanique rationelle*—whose apotheosis was reached in Lagrange's *Mécanique analytique* (1788). Most English mathematicians could no more have understood this work than they could have envisaged the theories of relativity.

It is difficult, if not impossible to know exactly how reform movements begin. The nadir of English science was possibly in 1800 when Thomas Young brought down a cataract of abuse on his head for daring to seem to question the opinions of Newton on the nature of light. Some ten years later a young Cambridge undergraduate named Charles Babbage who had already learned a little about Continental mathematics from a reading of Woodhouse's treatise reported that:

> In 1811, during the war, it was very difficult to procure foreign books. I had heard of the great work of Lacroix on the 'Differential and Integral Calculus' and which I longed to possess and being misinformed that its price was two guineas I resolved to purchase it in London on my way to Cambridge.

Babbage eventually bought his copy of Lacroix and became a convert to the Continental notation. He joined with likeminded scholars to found an 'Analytical Society' whose aim was the reform of English mathematics and the translation of the shortened version of Lacroix's work. Whatever the importance of this small but distinguished society, the reform of Cambridge mathematics was taken in hand by a group of young men that included George Peacock, later Dean of Ely, John Herschel, the son of Sir William Herschel and later himself a famous astronomer, Babbage, Ivory and William Whewell. Of these men it was probably Whewell who did more for mathematics and certainly for the reform of engineering mathematics, than anyone else. In the following decade Whewell started publishing books on the new mathematics and on mathematics for engineers. The days of Emerson, Banks and Olinthus Gregory were at last over.

Babbage, on the other hand, is nowadays best known for his contributions to computer technology. Simple calculating machines had been known for a very long time and when in 1812 Babbage invented his 'difference engine', as he called it, for mechanising the computation of mathematical tables he was not apparently making a revolutionary contribution to the art. In 1833, however, he conceived his 'analytical engine' which represented a great advance. This machine, as the name implies, could have handled complex algebraic formulae and carried out a virtually unlimited range of calculations. The natures of the formulae were to be imposed on the machine by means of punched cards; an idea being taken directly from the Jacquard loom that had first appeared in England in 1816. The analytical engine would have been the first digital computer, but it was before its time and although it cost Babbage a great deal of his own—and the country's—money, it was not completed. It is likely that his unhappy experiences with these machines made Babbage pessimistic about the present state and future prospects for science in the England of his day.

Whatever doubts there were about science—and they were so considerable as to lead eventually to the establishment of the British Association for the Advancement of Science (1830) —there were few if any about the healthy state of technology in this country. This satisfaction was possibly unfortunate and it may have contributed to the slow and reluctant development of higher technological education. The argument that one should not interfere with something that is eminently successful is very hard to counter; regrettably, by the time that one wakes up to the fact that it is no longer successful, it is usually too late to do anything.

At one point and at one level at any rate there was dissatisfaction and something effective was done. This was the matter of working-class education and more particularly the education in science and mathematics that was felt to be suitable for those engaged in the new industries created by the industrial revolution. In the early 1820s a most remarkable movement spread rapidly across the whole of Britain: the establishment of Mechanics' Institutes. These were intended to provide working men with evening instruction, after work, on the scientific principles underlying the technical processes of the industries that employed them. In this way it was felt workmen would be given an added interest in their work and, at the same time, the process of innovation and improvement be duly accelerated. The precise intentions of the founders, who represented a wide cross-section of the community, were never clearly formulated but of the initial success of the movement there can be no doubt whatever.

The first Mechanics' Institute was in Glasgow, founded by Dr George Birkbeck, a Quaker physician, and this was soon followed by a similar institution in London; the latter eventually developed into the present Birkbeck College. The greatest concentration of Mechanics' Institutes was, however, in the West Riding of Yorkshire, South Lancashire, North Cheshire and Derbyshire; the largest institutes being in Liverpool, Manchester, Huddersfield, Leeds and Halifax. There were smaller concentrations in the mining areas and round the ports. The Black Country area and Birmingham were curiously slow to establish Mechanics' Institutes; not until 1851 was there a tolerably successful institute in Birmingham.

There are two tentative conclusions we can draw from the Mechanics' Institute movement. In the first place it confirms our supposition that there was a numerous class of technicians in England who were, without being sophisticated technologists, quite capable of appreciating the value of scientific methods and new scientific knowledge in industrial processes. In the second place, the main centres of the movement were the textile areas where the pressures of innovation had been particularly strong and where, one infers, the concentrations of technical talent had been most marked. The implication was that by the 'chain effect' new technologies and new industries had grown up to service the great textile industry.

The problem of the power loom was really one of solving a number of individual details rather than of making an entirely new machine. Edmund Cartwright had been the pioneer in the eighteenth century, but real success was only achieved gradually as engineering standards rose. Richard Roberts, the machine tool engineer, produced the first satisfactory power-loom in 1822 and as he went on to produce an automatic self-acting mule in 1825 he may be said to have completed the final steps to the mechanisation of the textile industries. In less than sixty years an industry that had been a cottage-based craft was transformed into a mechanised mass-production industry. The refinement of textile machinery coincided at this stage with the progressive improvement of the steam-engine and the steam-powered cotton mill became the characteristic production unit in the industry. Nineteenth-century industrial England was being established.

However, the period 1825–1851 was in contrast to what had gone before, pedestrian and respectable: in some respects curiously like the second quarter of the eighteenth century. It was a time of consolidation, cautious reform and progressive evolutionary improvement. With one notable exception there were no really significant innovations or advances: the end of the period was much like the beginning with virtually everything very much better but little that was really new. On the debit side France had lost her way and the great period of French creativity, in science and technology, ended in the years after 1825. Why this should have happened is once again a subject of great conjectural interest. Perhaps the losses of the great war produced an exhaustion that entailed a low level of creativity; perhaps the orthodoxy of the immensely successful Ecole Polytechnique inhibited change; perhaps other countries were beginning to overshadow France. Germany in particular had been showing signs of a scientific awakening from about 1810 onwards. An association of scientists was established and was soon taken as a model for a similar body in Britain; in 1825 a chemical research laboratory, which rapidly gained an international reputation, was opened at the University of

Giessen under the energetic direction of Justus Liebig. But whatever the reasons for France's decline and however promising the omens in Germany, there can be no doubt that the middle period of the nineteenth century marked Britain's ascendancy in science and her maintenance of a comfortable lead in technology. Only towards the end of the period did certain rather ominous signs that all was not running smoothly begin to appear.

In 1832 Charles Babbage, then Lucasian Professor of Mathematics at Cambridge—Newton's professorship—applied his versatile and inquiring mind to a study of the new technological society that had developed in England. His book, *On the Economy of Machinery and Manufactures* was the second general study he wrote on the theme of science, technology and society: the first had been his *Reflections on the Decline of Science in England* (1830). The work that had contributed towards the foundation of the British Association, in imitation of the *Deutsches Naturforschers Versammlung*, *The Economy of Machinery* is a remarkably perceptive book in which Babbage shows a high degree of technological understanding and foresight. He even looks forward to the application of the division of labour in science, and to applied science as it developed later in the nineteenth century. He had read his Lean reports and was fully aware of the benefits of pooling technological information. He even prepared a questionnaire, remarkably modern in style, by which an inquirer could assess the technological effectiveness of any particular factory. But perhaps one of his most interesting observations related to obsolescence:

> Machinery for producing any commodity in great demand seldom actually wears out; new improvements by which the same operations can be executed either more quickly or better, generally superseding it long before that period arrives; indeed to make such an improved machine profitable, it is usually reckoned that in five years it ought to have paid for itself and in ten to be superseded by a better.
>
> 'A certain manufacturer' says one of the witnesses before a committee of the House of Commons 'who left Manchester seven years ago, would be driven out of the market by the men who are now living in it, provided his knowledge had not kept pace with those who have been during that time constantly

profiting by the progressive improvements that have taken place in that period'.

The pace of innovation in certain industrial areas of Britain was obviously very hot. It is clear, too, that a class of machinery-makers had grown up who were responsible for innovation and who thereby put pressure on the manufacturers. Some light was thrown on this by Andrew Ure in a book, *The Philosophy of Manufactures* (1835) that was a fairly obvious imitation of Babbage's work. Ure's book is less discursive than Babbage's and his concern was almost entirely with the textile industries. Nevertheless as the textile industries were associated with a wide range of other industries and technologies this was not such a drawback as it might appear. The frontispiece to this book represented the weaving shed of a cotton mill at Stockport. The artist evidently exercised his licence and exaggerated the scale of the building: nevertheless the rows of identical power looms stretching interminably into the distance, like a vast mechanical army on parade, clearly represented something new in manufacture: the mass production of complex machines. This was a distinct step towards that very characteristic industry of the twentieth century: the mass production of motor cars. Ure makes the point very clearly:

> Indeed the concentration of mechanical talent and activity in the districts of Manchester and Leeds is indescribable by the pen and must be studied confidentially behind the scenes in order to be duly understood or appreciated.
>
> The following anecdote will illustrate this position. A manufacturer at Stockport . . . being about to mount two hundred power looms in his mill, fancied he might save a pound sterling in the price of each, by having them made by a neighbouring machine maker instead of obtaining them from Messrs. Sharp, Roberts in Manchester, the principle constructors of power looms. [He] surreptitiously procured iron patterns cast from one of the looms of that company which in its perfect state cost no more than £9 15s. His two hundred looms were accordingly constructed at Stockport, supposed to be fac-similes of those regularly made in Manchester and they were set to work. Hardly a day passed without one part or another breaking down, insomuch that the crank or tappet wheels had to be replaced three times, in almost every loom, in the course of twelve months . . .

In the end the foolish and greedy manufacturer was forced to go to Sharp, Roberts and order the looms he was unable to make for himself. The specialised machine makers, capable of manufacturing in large batches, or in mass producing, could surpass the individual who wanted to make his own machines. Machine tools, specialised and of high precision, enabled them to do this. Of Sharp, Roberts, Ure remarks:

> Where many counterparts or similar pieces enter into spinning apparatus they are all made so perfectly identical in form and size by the self-acting tools, such as the planing and keygroove cutting machines, that any one of them will at once fit into the position of its fellows in the general frame.

The development of such machine tools enabled such a high standard to be achieved that parts were interchangeable between machines; accordingly mass-production became feasible.

From the social point of view these developments imply the existence of a numerous class of highly skilled technicians without whose work the automatic machines, central to the industry of the time, could not possibly have been made. It is an instructive, and can be a moving experience, to turn over the pages of the pattern-books and engineering drawings; the specifications and order books of the textile and engineering firms of a hundred and more years ago. Each item reveals the high degree of skill and intelligence of the numerous people who made it all possible. Our present prosperity, indeed the prosperity of all the wealthier communities of the world, was built up on the varied skills of these people, now long dead. Conversely, the problem of the undeveloped nations of today is, in part at least, that of creating the same sort of classes of skilled technicians as those that grew up in the industrial areas of Britain during the latter part of the eighteenth and the first half of the nineteenth centuries; men who are known to us through the records they left behind, through the Mechanics' Institutes they attended (their universities) and through those old machines that still survive in museums and similar institutions.

As we remarked, however, the age was a rather dull one, of assured progress, evolutionary improvement and with few startling innovations. The great exception was Michael Faraday's discovery of 'electromagnetic induction'. Oersted's discovery that an electric current deflected a magnetic compass needle placed under it, and the converse discovery, made soon afterwards in France, that a magnet would deflect a moveable wire in which an electric current was flowing, prompted a number of people to try to find if one electric current would cause another to flow in a neighbouring, but quite separate loop of wire, or circuit. This could be expected on analogical grounds, for it was well known that a magnet would magnetise a nearby piece of soft iron, while an electric charge will give rise to, or 'induce', a charge on a nearby body. We should therefore expect to find that an electric current, which has strong affinities with magnetism and with static electricity (one can collect it in a Leyden jar or condenser) can also, in some way, induce a sympathetic current in a nearby circuit. For ten years scientists tried to find this expected effect, but although they included men of the calibre of Ampère they failed. Then, in 1831 Michael Faraday, working at the Royal Institution in London, found the answer. In the course of very careful experiments Faraday noticed that a current flowing in a wire had an effect on a nearby circuit only at the very instant when it was being switched on. Then, for a very brief moment, a current flowed, in the opposite direction to the first current, in the nearby or secondary circuit. And when the current was switched off in the first, or primary circuit there was another momentary current in the secondary circuit, this time in the opposite direction to the first induced current. With great genius Faraday systematically explored this, apparently trivial, phenomenon that could, on the face of it, have been due to a host of accidental factors.* He showed that the true law of induction was that a *changing* current induces a *changing* current; a steady current does not induce a steady current. This explained why Ampère and the others had failed: the effect they expected and sought did not exist. But Faraday went on to show that he could obtain brief induced currents without using a primary current at all; moving a small permanent magnet

---

* Thus the effect might possibly have been due to mechanical vibrations caused by the acts of switching the current on and off; or it might have been due to some sort of leakage of electricity through the air.

very rapidly near his secondary circuit produced exactly the same effect. In this case the induced current persisted for only so long as the magnet was actually moving. It was not, therefore, the change of current *per se* that had caused the induced currents in the first experiments; it was the changing magnetism caused by the changing primary currents.

All this was immensely important; it marked the beginning of what is known as 'field theory' and it started a revolution in physical science. Let us consider, very briefly, two points that have emerged so far. In the first place Oersted, although he did not realise the implications, had discovered something quite new in science and human experience. Men knew of simple attractive forces such as gravity—the story of Newton and the apple need not be repeated—and of correlative repulsive forces such as that between the north poles of two permanent magnets. But an agency that, acting over a distance, through air or empty space, could deflect something was an entirely new experience.

In the second place we must recapitulate Faraday's ideas about 'lines of force'. If we place a sheet of clean white paper on top of a permanent magnet, sprinkle iron filings on the paper and then give it a gentle tap, the iron filings arrange themselves into curved 'lines of force' running between the two poles of the magnet. The same sort of lines of force can be obtained if instead of a magnet, we use a wire carrying an electric current and running vertically, at right angles, through the paper. In this case the lines of force take the form of concentric circles with the point at which the wire passes through the paper as the common centre. Now for Michael Faraday these lines of force really existed in space; they constituted a magnetic 'field' and it was the changing magnetic field that induced momentary currents in a secondary circuit. We shall discuss the consequences of these views of Michael Faraday's later on (see below, p. 173).

The first practical consequence of Faraday's discovery of electromagnetic induction was that it made possible the invention of the dynamo as a generator of electricity. Evidently if a loop of wire, or circuit, is moved rapidly through a magnetic field by a steam engine or other source of power an electric current will flow in it. The magnetic field can be provided either by a permanent magnet or, much better, by another current flowing in an electromagnet. After 1831, therefore, it was possible to generate electric currents mechanically. And, we must add, the electric motor and the transformer, as well as the dynamo, were feasible. The electromagnet had been invented by William Sturgeon as early as 1825, so that the foundations of the electric power industry had been laid, as far as science was concerned, before the end of the first third of the nineteenth century.

However, before an electrical industry could spring up several important obstacles had to be overcome. There was the competition from steam engines, growing more efficient every day, and competition from gas lighting, becoming cheaper and more widespread all the time. At the same time, before electric power could come into its own, a large number of ancillary inventions had to be made and a manufacturing industry established. All this was in the end achieved, but it necessarily took a rather long time. The first uses for electricity were, as we remarked before, to provide telegraphy, mainly in connection with the developing railway services, and to establish the electro-plating industry. Telegraphy and metal-plating were two things that could be done much more effectively and cheaply by electricity than by any other method at that time.

THE MID-CENTURY: THE PINNACLE
OF BRITISH ACHIEVEMENT

After the quiet decades the middle of the nineteenth century was, as far as science was concerned, a great period of synthesis. Maxwell's theory, based on Faraday's pioneering researches established the electromagnetic nature of light and radiant heat and at the same time predicted the existence of other electromagnetic waves of much lower frequency that might be set up by simple electrical oscillations. The doctrine of the conservation of energy was put forward mainly as a result of Joule's researches, and thermodynamics, Carnot's science, was established on the sound foundation of the energy doctrine, as Joule had established that heat was a form of energy. In many respects the two great advances at this time—energy and field

F

theory—represented the supersession of the old Newtonian natural philosophy and the substitution of a new and more general science: physics. Apart from physics we had Mendeleeff's great systematisation of chemistry, the periodic system, and in biology, Darwin's theory of evolution by natural selection. As has been remarked one could hardly be blamed for feeling that synthesis was in the air.

Socially and technologically the age was interesting for the sequence of great exhibitions that were held following the success, as far as England was concerned, of the first great International Exhibition of 1851. On that occasion there could be no doubt whatsoever: England was the leading industrial nation of the world and an international jury merely confirmed the fact when they awarded most of the prizes to Britain.

From the historian's standpoint the most revealing of the displays, mounted by other nations, was that of the United States. The impression given was that America was a newly settled, pioneering country: the main exhibits being things like agricultural implements—C. H. M'Cormick's Virginia grain reaper was particularly admired—saddles, guns—Colonel Colt's revolver was prominently featured—mineral and vegetable products, wood-working machines and small-town craft manufactures such as examples of the book-binders' and printers' arts. For the rest the American display included some of Ericsson's machines, a floating church for the particular benefit of sailors, Morey's sewing machine and the now-forgotten inventions of Henry Pinkus. There was, in short, little hint of the mighty industrial and technological power to be developed in the next hundred years; in 1851 British technology was supreme in the old and in the new worlds.

The expression 'machine tool' had still not come into common use at that time; things like lathes and boring machines were classified as manufacturing machines and tools. Nevertheless there was general recognition of the basic industrial importance of these machines. Babbage, observant and fluent as ever, summed the matter up when he commented, in his book *The Exposition of 1851*, '. . . it is not a bad definition of *man* to describe him as a *tool making animal*'. One need not accept this definition *in toto* to concede that, in the context of his essay, Babbage had made a good point. By the time of the exhibitions of 1851 and 1862 a wide range of industrial machine tools had become available. London, Manchester, Leeds* and Glasgow were clearly established as the centres of the machine-tool manufacturing industry. There were very few machine tool firms in Birmingham, Coventry and the Black Country; an indication, surely, that in days before the bicycle and the motor-car, machine tools were particularly associated with textile machinery making and with heavy (steam) engineering. The industrial lathe ranged in size from the small, instrument makers' machine to monsters that could turn major components for locomotives and marine engines. The planing machine that Richard Roberts had been the first, or at any rate one of the first, to introduce in 1817 and the slotting machine, developed by Fairbairn and Lillie in 1828, were both in common use and so were pillar drills and multi-spindle drills. Milling machines were beginning to appear, while shaping machines and a wide variety of special machine tools such as rivetting machines, file cutting and screw cutting machines, were being made. It was even possible to discern future trends towards the automatic and the profiling lathe.

Important newcomers in 1851 were the radial drill and the steam hammer. The former consists essentially of a strong cylindrical pillar, up and down which a horizontal arm can be moved and fixed at any desired height. This arm can also be turned horizontally to point in any direction, apart from a small sector. Mounted on the arm is a moveable carriage that supports the vertical drill spindle. Power is transmitted by belting from line shafting in the ceiling to a set of pulleys, inside the closed sector at the base of the pillar, and thence by bevel gears and shafting to the spindle. By virtue of its degrees of freedom of motion and adjustment the radial drill is a very flexible tool.

The steam hammer was invented by James Nasmyth in response to the need for a method of forging the very large paddle shafts required for the big steam ships of the period. Traditionally forging had been done by means of the tilt hammer, a tool of great antiquity and used also in fulling stocks and the early paper industry. But tilt hammers could not take

* The industry subsequently deserted Leeds for the textile towns of Huddersfield, Halifax and Bradford.

the big paddle shafts and a new tool was required. Nasmyth therefore developed the steam hammer, a very simple but effective machine that consists of a vertical cylinder fitted with a piston connected to a heavy hammer head. High pressure steam, applied under the piston raises the hammer which can then be allowed to fall without check, or, by controlling the efflux of steam, at any desired speed and stopped at any chosen point. The sheer size and power together with the precision of the steam hammer greatly impressed the early Victorians. For them it epitomised modern man's complete control over forces so much greater than simple muscular power.

The doyen, publicly acknowledged, of British machine tool makers at this time was Sir Joseph Whitworth. A systematic Smeatonian-type improver Whitworth had, in fact, carried the industry forward, as Maudslay had done at the beginning of the century by raising the standards of manufacture and therefore of tool performance to new heights of accuracy. He improved the design of his tools and achieved the maximum rigidity consistent with a given weight by using hollow castings. As the official handbook to the 1862 exhibition put it:

> . . . in a given weight, in framing exposed to transverse strains all the material must be put as far as possible from the neutral axis—i.e. the framing must, like the human bone, be hollow.

This was the time, we remember, of the superb Britannia tubular bridge across the Menai straits in the design and construction of which two Manchester engineers, William Fairbairn and Eaton Hodgkinson had collaborated with Robert Stephenson. Galileo's pioneer studies of the strength of materials had come a very long way indeed.

Whitworth also won distinction by the accuracy of his gauges, dividing engines and micrometer.* So uniformly excellent was his work that it was clearly recognised at the time that he could not be excelled. But later on a post-Whitworth complacency set

---

* Robert Mallet, writing in the catalogue of the 1862 exhibition could not approve of Whitworth's micrometer. How, he asked very reasonably, could it be accurate when the sense of touch cannot be magnified? Optical methods permit magnification and must therefore be superior for accurate work. Mallet was, of course, quite correct; ultimate accuracy must be sought by optical means, but this did not make Whitworth's micrometer useless, or even inaccurate; far from it.

in and the English machine tool industry lost its lead to other countries: to America and then to Germany as well. Indications that this was happening appeared fairly soon. In the official Report on Machine Tools—the expression was now in common use—at the Paris exhibition of 1867, J. Anderson remarked that Whitworth's style was now being copied successfully in France and Germany, who were catching up on England. But America was offering something new: exquisite workmanship together with ingenious new designs. We, the Report added, need some fresh thinking.

Much the same trend had appeared in another industry. British agriculture, then as now a highly efficient industry, was served, to judge by the 1851 and 1862 catalogues, by an excellent range of implements—ploughs, harrows, rakes, rollers or clod-crushers, threshing machines, and steam traction engines —and these continued in use with surprisingly little modification, through the long decades of depression from the 1870s up to 1939 when the revival of the industry began. There was (unfortunately) no American display at the 1862 exhibition, but in 1851, as we saw, the M'Cormick reaper had aroused great interest. This machine, with its horizontal cutting teeth and its very distinctive paddle wheel that bent the stalks into the teeth, was the ancestor of the line of reaping machines that has culminated in the combine harvester.

More modest, but no less important, was the debut of the turret and capstan lathes that were developed in America some time between 1840 and 1850. Traditionally the invention of the turret lathe, the larger of the two, is ascribed to the tremendous demand for personal firearms following the rush to the West and the popularisation of the Colt revolver. Certainly the turret lathe was pre-eminently adapted for the production of very large numbers of identical components; it enabled several different turning operations to be carried out rapidly, one after the other, without requiring the intervention of a skilled tool-setter before each operation could commence. This was done by having all the necessary cutting tools mounted in such a way that they could be brought into action, one after the other, by a revolving holder. In some of the early machines the tool holder consisted of a horizontal drum with all the tools mounted peripherally on the face opposite the chuck. But the turret form proved superior. This has, in place of the slide rest, a large octagonal vertical turret, one of whose faces is always at right angles to the axis of the spindle. Eight cutting tools can be bolted to the faces of the turret and brought into use as required merely by turning the turret on its axis. In this way a workpiece can be drilled, bored, turned and screw-threads cut with only the simplest movements on the part of the operator. Once a skilled tool-setter has fixed the cutting tools on the turret an unskilled operator can make thousands of identical components easily and quickly: bathroom tap fittings as easily as parts of revolvers. It seems probable that, unless they were exhibited under the title of 'screw cutting lathes', no British manufacturer displayed a turret or capstan lathe at the 1862 exhibition.

TECHNOLOGY AND COSMOLOGY

The availability of cheap and abundant power was vital for the industrial progress of the western nations. In the middle of the nineteenth century this meant, simply enough, the reciprocating, expansively operated steam-engine, then in a phase of intensive evolutionary improvement. The full potential of the designs pioneered at the beginning of the century were being progressively realised. Even the compound engine, advocated by Hornblower and then by Woolf, took on a third lease of life when the McNaught compound engine was produced in the middle of the 1850s.

Before, however, we go on to consider the question of power and more particularly the ultimately successful rival to the steam-engine, the air-engine, we shall discuss the destiny of Sadi Carnot's ideas.

Shortly before the Exhibition of 1851, James Prescott Joule, a Manchester scientist who combined devotion to science with experimental gifts and philosophical perception amounting to genius, showed that no matter what processes or substances are involved the exchange rate between mechanical work and heat is constant. He demonstrated, in other words, that heat and mechanical energy are interchangeable; and he measured the new constant of nature, the mechanical equivalent of heat, with astonishing precision. This meant the rejection of both the 'caloric' theory and of the basic axiom, universally accepted until then, that heat is always and under all circumstances conserved. In its place Joule substituted the axiom that energy is always conserved.

As R. J. E. Clausius pointed out, Joule's work meant that Carnot's theory of the heat-engine had to be modified. Although Carnot had not realised that the work done by an engine was at the expense of heat from the furnace, that is heat actually transformed into mechanical energy, he was not wrong in supposing that there *must be* a transfer of heat from a hot to a cold body whenever an engine is worked: some sort of condenser is no less essential than a furnace and boiler.

We can no longer use Carnot's proof that no engine can be more efficient than a reversible one, since the simple hydraulic analogy has now broken down. A column-of-water engine cannot harness its own tail-race; it cannot be driven by water at a lower level than itself. But there is nothing, on the face of it, absurd in supposing that a heat-engine might transform some of the heat energy from the cold body, as well as from the hot body, into mechanical energy and that such an engine might be more efficient than a Carnot (reversible) engine. Why, indeed, should we not be able to tap the immense reservoir of heat energy contained in the Atlantic ocean? An engine that could do this would achieve, not quite perpetual motion, but something almost as good: a virtually limitless supply of energy.

All our experiences of heat-engines, however, convince us that this is not possible; that, as Carnot had insisted, the passage of *some* heat from a hot to a cold body is entirely necessary if mechanical energy is to be obtained. Clausius therefore proved, very simply, that if an engine can be made which is more efficient than a reversible engine working between a hot body and a cold body then if the more efficient engine drives the reversible one backwards the net effect would be, for the combination, merely a transfer of heat energy from the cold body to the hot body. This, Clausius argued, is opposed to all experience, not merely of heat-engines, ideal or otherwise, but of all the phenomena of heat, of all the energy transformations in nature. In other words Clausius postulated that among all the transformations of energy that scientists study, or will ever study, no matter how complex or remote, no instance will ever be found of low temperature heat energy being transformed into high temperature heat energy without some sort of compensating change taking place. As he put it: 'It is impossible for heat to pass from a cold body to a hot body without compensation'. This is known as the Second Law of Thermodynamics.* It is entirely general and its scope extends far beyond our simple, everyday experiences that hot bodies tend to cool down and that cold bodies never, of their own accord, warm up.

Clausius, like his contemporary William Thomson (Lord Kelvin) and like Carnot earlier on, regarded heat as a motive agent of cosmological significance. He, however, was confronted not with simple problems of the passage of heat, regarded as some special type of fluid, but with transformations of heat and mechanical energy. In principle all the transformations of energy are of the same sort, or are equivalent to one another, and we must be careful not to let old-fashioned expressions like 'the flow of heat' or 'heat capacity'† fool us into thinking that heat is, in some way, a distinct, individual substance.

Kelvin, thinking along parallel lines to Clausius although not so profoundly, had seen that heat, considered as a form of energy, is in one sense peculiar. It is that form of energy into which all the other forms tend to transform themselves; and, furthermore, the final destiny of all forms of energy seems to be heat at the lowest available temperature. There is, Kelvin remarked, a universal tendency in nature to the dissipation of energy. The most efficient steam-engine wastes mechanical energy in (unavoidably) generating heat by friction. This heat energy tends towards the coldest body in the neighbourhood and so becomes less readily available as a source of useful work: it has been dissipated. In fact all the energy in the universe is slowly, inexorably transforming itself to the level of

* The First Law of Thermodynamics expresses the exact relationship between heat and mechanical energy.

† The expression to which Clausius took particular exception was 'latent heat'. He held that it implied a 'heat substance' that could be hidden away! Only now, a hundred years after Clausius' papers, is 'latent heat' passing out of use.

the most diffused, least concentrated heat energy; even though no energy is actually being destroyed.

Energy transformations in nature range from the simple phenomenon we call the conduction of heat (in other words, its transformation from a high temperature state to a low temperature one), through a multitude of physical, chemical, biological processes to the behaviour of man-made heat-engines. Marking the limit in this direction is the reformed Carnot engine which under given conditions can transform the maximum amount of heat energy into mechanical energy; or, as only Clausius had seen, can produce the maximum amount of mechanical energy from the passage—or transformation—of heat energy from the hot boiler to the cold condenser.* With precise scientific instinct Clausius then set out to find a suitable mathematical expression to equate these different transformations. The conversion of a given amount of heat energy at a given temperature into mechanical energy is equivalent to, or can be replaced by, the passage of a second amount of heat energy between two given temperatures, or the passage of a second amount of heat energy between yet another pair of temperatures, or . . . but one need not go on. The problem was, for Clausius, to find a common numerical value to express the fact that all these transformations are equivalent. In fact this was the same sort of situation as that which confronted Newton when he set out to quantify, or mathematicise, the mechanism of the heavens.

After some difficulty Clausius found an expression that he called the 'equivalence value' of a transformation and that could be computed for transformations of heat into mechanical energy or *vice-versa*, or the passage of heat energy from one body to another. The equivalence value amounted to the quantity of energy transformed divided by the temperature (absolute) at which the transformation took place.† If the transformation

* Since the heat energy that is not transformed into work *must* be transmitted to the cold body, or condenser, it follows that these two cases mutually imply each other. In other words, we can say that the work performed by a Carnot engine is determined *either* by the conversion of a certain amount of heat energy into work (Joule) *or* by the transmission of the rest of the heat energy to the condenser (reformed Carnot).

† The equivalence value of the passage of heat energy between two different temperatures is easily determined. Let us suppose that the heat is

was in the direction favoured by nature—the passage of heat
energy from a hot to a cold body, or the conversion of mechanical energy into heat—the equivalence value was given a
positive sign; if the transformation was in the opposite direction
it was given a negative sign. In the case of a perfect Carnot
engine the sum of the equivalence values of the transformations
in one complete cycle would be zero, for they cancel out. But
in the case of all other—that is, real—engines, and of natural
processes generally, the sum is always a positive quantity; in
other words, the net effect of all the transformations is equivalent to the passage of heat energy from a hot to a cold body, or
to the conversion of a certain amount of mechanical energy into
heat. This accords with Kelvin's observation that there is a
universal tendency towards the dissipation of energy.

Clausius then went on to apply his theorem of the 'equivalence of transformations', as he called it, to the internal transformations that take place when a body changes its condition
or state. If a certain amount of heat energy enters a body—a
gas or vapour for instance—then usually its pressure and
volume increase, the temperature rises and work is done on the
container or surroundings as the body expands. The rise in
temperature, Clausius argues, indicates an increase in the
'internal' heat (that is, the kinetic energy of the constituent
molecules), and the expansion of the body means that the
separation of the molecules has become greater. These are the
two internal effects of the entry of heat energy into a body, and
the second one is the corollary of the external work the body
does in expanding. The equivalence values of the two transformations are obtained, says Clausius, by dividing the in-
crement in the internal heat and a measure of the molecular
separation by the absolute temperature. The sum of these two
equivalence values he calls the 'transformation content', or the
entropy, of the body. In the case of a Carnot engine the entropy
changes of the working substance—air or steam—are of equal
size but opposite sign to the equivalence values of the external

---

transformed into mechanical energy at the higher temperature and the
latter reconverted back into heat at the lower temperature; the difference
between the equivalence values of these two transformations is the equivalence value of the passage of heat between the two given temperatures.

transformations so that, after one complete cycle by such an engine, the net entropy change of the working substance is zero.

The change in entropy of a body indicates the extent to which, and the direction in which the body changes its energy state. If, after a series of changes, simple or complex, the entropy has increased, then the net effect, or the total equivalence value of all the changes must be the same as if a certain amount of heat had entered the body at a certain temperature, or the body had expanded at that temperature. To the extent that a body expands or warms up or does both its entropy increases; to the extent that it cools down, or contracts or does both its entropy decreases. If, for example, a body expands and at the same time loses heat energy (so that it cools down) to exactly the same equivalence value, its entropy change is zero. In the very special case of a gas expanding into a vacuum without doing any work (what would there be for it to press on?), its entropy will still increase even though it does not absorb any heat energy from outside (why should it? no work has been done). In this case the entropy has increased because the gas has expanded. If we want to restore the gas to its original condition we must compress it, but in order to prevent its temperature rising we must allow the heat energy, generated by compression, to pass into another body at a lower, if only slightly lower, temperature. The entropy of this body must increase as it acquires the additional heat energy; indeed, because the divisor (the temperature) is lower, if only slightly lower, its entropy will increase more than the entropy of the compressed gas is reduced.

We cannot, in practice, deal with bodies in isolation. Everything that cools down without performing work must give up heat energy to a body, or bodies, that warm up. And since the temperature is the divisor, or denominator, in the measure of entropy change, it must follow that the reduction in entropy of the first body *must* be less than the increase in entropy of the second. In fact, in any closed system of bodies interacting on one another, the sum total of the entropy changes must always be a positive quantity.* This led Clausius to make a very

---

* Were it otherwise the unavoidable implication would be that heat energy must have passed, without any sort of compensating change, from

famous generalisation: 'The energy of the Universe is constant; the entropy is always tending to a maximum.'

This was the ultimate statement of the cosmology of heat. It was in consequence of the achievements of medieval technics that the universe could, in the seventeenth century, be likened to the work of a cosmic Clockmaker; magnificent, but not yet perfect for the Clockmaker (God) had to intervene from time to time to keep it running smoothly. By the early nineteenth century Laplace could view the universe as a perfectly balanced dynamical system, requiring no supervision; no God at all. In the middle of the nineteenth century the picture changes again and the universe, interpreted through thermodynamics, is seen as a somewhat-less-than-perfect heat-engine that must eventually dissipate all its energy concentrations and end in timeless, eternal stagnation.

It is, in one way, a pity that Clausius' division of entropy into two components has been almost totally forgotten for it surely makes the origin and nature of this—elusive!—concept easier to understand. It cannot be explained away analogically and to define it, as many text-books do, as a property or function that increases when a body absorbs heat energy is, although correct, both insufficient and, to the candid reader, arbitrary. The best approach would seem to be historical: through the ideas and opinions of the man who formulated the concept.

In the event, however, the simple, undivided concept of entropy proved entirely adequate for all purposes and its application, in chemistry and physics as well as in engineering has been wide and fundamental. Entropy change indicates how energy is distributed in different processes and how equilibrium states are defined; and it does so without the use of hypothetical or sub-microscopic entities.

Thus the endeavour that began with James Watt's search for the maximum 'duty' to be obtained by a simple process (burning good quality coal) was generalised during the following hundred years to cover the transformational processes of nature. And reflections on the heat engine, between 1824 and

---

a colder to a hotter body. And this contradicts the Second Law of Thermodynamics. We can say, therefore, that this statement is an alternative expression of the Second Law.

1865, resulted in a profound change in man's view of his universe and its destiny. For the third time technology had had a direct and major impact on the world of thought and speculation.

COSMOLOGY AND TECHNOLOGY

If the technology of the heat-engine profoundly affected cosmological ideas, certain cosmological views seem to have had equally marked effects on technology. The middle of the nineteenth century was the time when, as far as Britain was concerned, faith in progress was at its strongest. The achievements of physical science, the formulation of Charles Darwin's evolutionary biology coupled with the facts of social, economic and technological advance combined to impress upon nearly everyone the reality of progress and its almost certain continuation. Even the least imaginative could hardly escape the implications of the great exhibitions and the rise in living standards brought about by such developments as that of the railway system. It was reasonable to believe in perfectible society, if not perfectible man, and thinkers like Herbert Spencer resolved to find the law of social progress; the sociological analogue of the biological doctrine of evolution.

Such a climate of opinion would, one supposes, have certainly affected technology; setting men's faith in it at a still higher level and inviting speculations about the even ampler prospects ahead. And yet, curiously enough, this does not seem to have happened. There seems to have been little or no appreciation of 'open-ended' technology. Today we regard human progress as a rather doubtful proposition—two world wars and the known potential of 'weapons technology' have dealt that simple faith some heavy blows—but we couple our pessimism about moral and social progress with almost unlimited faith in the possibilities of technology. Nothing is beyond the reach of technology; beyond the amazing powers of the organised applied science that has been developed in the present century. In short we seem in the course of a hundred years to have reversed the Victorian attitudes to progress and to technology.

Mid-Victorian technology was boxed in, as it were, by four powerful factors that necessarily limited men's expectations of it. Firstly, there were the firmly established laws of science. In the second half of the nineteenth century it came to be believed that the physical sciences were nearly complete. Rational mechanics and recently electromagnetism and optics were near-perfect sciences while Daltonian chemistry, based on the concept of the immutable atom, was set in the frame-work of Mendeleeff's periodic system in which there was a place for all the elements, known and shortly to be discovered. Secondly, there was the establishment of thermodynamics that fixed definite, unalterable limits on what could and what could not be done by heat-engines. Thirdly, the new science of geology coupled with the near completion of the long process of geographical discovery indicated clearly enough the nature of world resources. There were no real secrets left to find, raw materials were much the same everywhere and although now and then some oddity might turn up in a remote part of the world it was not likely to affect the lives of the vast majority of men. Finally there were the 'iron' laws of classical economics. Having established world resources and discovered the laws that determine, in physical and chemical processes, the transformations of those resources we find, in the last resort, that we are circumscribed in all directions and any dreams of wonders yet undiscovered are subject always to the cold correction of scientific disproof. No engine or means of power more efficient than the steam-engine can be discovered; and such engines must always be limited by the cost of the fuel—coal—available in the particular locality.* Chemistry and physics hold out no hope of any other or superior fuel or mode of power. Only gradual evolutionary progress is possible. Paradoxically therefore we conclude that at the time when the idea of progress was most firmly held, in that sector of human endeavour where progress had most obviously and incontrovertibly taken place there were good reasons for supposing that future progress could only be of a pedestrian, evolutionary nature.

* I am very grateful to Dr A. J. Pacey for calling my attention to W. S. Jevons' observations on this point (see *The Coal Question*, London 1862).

THE HEAT-ENGINE

During the period when physical science seemed to be approaching completion and, more particularly, when the laws of thermodynamics were being worked out by Clausius, Rankine and Kelvin, progress in heat-engine technology was either evolutionary in the case of steam-engines or hesitant and amateurish in the case of air-engines.

There were exceptions. Joule and Rankine, for example, carried out studies of the air-engine. But air is a very bad conductor of heat and therefore difficult to heat up; furthermore its expansion is so slight that if reasonable power was to be obtained within a moderate temperature range enormous cylinders had to be used. The cylinders of Ericsson's engine, for example, were no less than fourteen feet in diameter so that although in theory it was quite successful, in practice it was not. The need for a small compact power unit had been felt for some time. In 1851 it was expressed clearly by Babbage and even if he came up with the wrong answer his argument was, in principle, correct:

> One of the inventions most important to a class of highly skilled workmen (engineers) would be a small motive power—ranging perhaps from the force of half a man [sic], to that of two horses, which might commence as well as cease its action at a moment's notice, requires no expense of time for its management and be of moderate price both in original cost and in daily expense. A small steam engine does not fulfil these conditions. In a town where water is supplied at high pressure, a cylinder and a portion of apparatus similar to that of a high pressure engine would fully answer the conditions, if the water could be supplied at moderate price. Such a source of power would in many cases be invaluable to men just rising from the class of journeyman to that of master. It might also be of great use to many small masters in various trades. If the cost per day were even somewhat greater than that of steam for an equal extent of power, it would yet be on the whole much cheaper because it would never consume power without doing work.

Babbage's suggestion then was the column-of-water engine. This was entirely reasonable in view of the fact that men like

W. G. Armstrong (later Lord Armstrong) were busy developing schemes for central pumped storage reservoirs to provide power, by means of high pressure mains, to drive dockside and warehouse cranes, and later to work lifts and elevators.* Remains of these systems are still to be seen. But the ultimate fate of the column-of-water engine in this country seems to have been to provide a discreet source of power to work the bellows of the organs in the large churches and chapels that were being put up in the growing industrial cities.

It is clear that the hot-air engine, in the form of the oil- or gas-engine, would easily have met Babbage's requirements and had they been at a reasonable state of development he would surely have mentioned them. The electric motor fills the bill too, but that required even more development and vast infusions of capital to provide for power stations and transmission systems.

An important factor at this stage was the rapid growth of the gas industry. In those towns where the rising journeymen and the small masters were to be found gas was being laid on for lighting, cooking and even domestic heating. Here then was a cheap 'self-stoking' fuel suitable for an external or an internal combustion air-engine and which, moreover, could provide immediate power when required while costing nothing when the engine was switched off. Accordingly, inventors continued to experiment with engines in which a mixture of air and gas was forced into the working cylinder under pressure and ignited by means of momentary exposure to a gas flame burning in a chamber adjacent to the cylinder: alternatively the mixture might be exposed to a rod of incandescent metal. But these experiments, while interesting, were unsystematic and apparently quite unrelated to the advanced theoretical knowledge available at the time.

In 1860 Lenoir produced the first commercially successful

* One great advantage pointed out by Armstrong was that such a power system would have no problems of 'peak loading'. A steam-engine working twenty-four hours a day, seven days a week, could keep the reservoir topped up although at peak load periods the demands made on the system would far exceed the capacity of the steam-engine. The principle of pumped storage has recently been re-introduced in the electricity supply industry where reservoirs are charged with water at night and the water let down through turbines coupled to dynamos when there is a peak demand.

gas-engine. In this engine a mixture of air and gas was drawn into the cylinder by the motion of the piston, the valves were then closed and the mixture ignited electrically; the exhaust gases being expelled from the cylinder on the return stroke. The engine was double-acting but it was also very inefficient. It was designed seemingly without reference to the scientific knowledge available and such insights as were apparent were derived from erroneous ideas. In fact, although the subsequent development of the internal combustion air-engine has more than justified the efforts of these early pioneers, they seem to have been well out of the main stream of power engineering at that time; a collection perhaps of dreamers, cranks and spare-time inventors. In view of the continued and apparently irresistible success of the steam-engine this was perhaps inevitable.

In 1862, however, a French engineer, Alphonse Beau de Rochas, wrote a short pamphlet in which the elementary tenets of thermodynamics were applied to the problem of the various heat-engines, including of course internal combustion engines. The pamphlet, which was in manuscript, was quite innocent of mathematics or of any other of the paraphernalia of advanced scholarship. Nevertheless Beau de Rochas laid down the important points briefly and clearly. He remarked that the grates of even the best steam-engines admitted far too much air for the really efficient combustion of the fuel and he went on to suggest that preliminary gasification of the fuel should allow of much more effective use. This led him to a brief discussion of a combined gas- and steam-engine, in which the waste heat of the first process provided steam for the second, and thence to the principles of the simple gas-engine.

Beau de Rochas referred to a paper by Victor Regnault (*Comptes Rendus*, 18 April 1853) concerned with the efficiency of heat-engines and containing an account of Carnot's principle. This, as it happened, was the only reference in the whole pamphlet and it can be assumed therefore that Carnot's arguments guided the subsequent discussion. Thus he considered heat-engines without preliminary compression of the air. Such machines will never be able to develop much power and he did not hold out great hopes for Lenoir's engine. Indeed such engines violate the basic conditions for the most efficient use of the elastic force of gases:

> These conditions are, in effect, four in number. 1st, the greatest possible volume of cylinder together with the minimum surface area; 2nd, the greatest possible speed of expansion; 3rd, the greatest possible expansion; 4th, the greatest possible pressure at the beginning of the expansion.

The only practicable way of approximating to these basic conditions is to carry out a four-stage operation:

> 1st, Air and gas are drawn into the cylinder by one complete stroke of the piston;
> 2nd, In the following, return, stroke the air is compressed;
> 3rd, Ignition at the dead-point and expansion in the following, third stroke;
> 4th, Expulsion of the burned gases from the cylinder during the third and last stroke.

This, of course, is the specification of the four-stroke cycle.

In 1876 Otto produced his four-stroke engine but without apparently having heard of Beau de Rochas and with some erratic ideas about the principles which made his engine efficient. This was not, of course, the first time the right design had been deduced from wrong principles! In any case the rate of progress was now accelerating. The carburettor had been invented and the gas- and oil-engine could advance together: indeed at the margins it is difficult to distinguish them. The small workshops in industrial towns now had their gas-engines and on large farms the agricultural machinery was now driven by oil-engines. Thus two demands which the steam-engine could not meet and which electric power was not yet sufficiently well developed to cope with, had been satisfied.

While this was going on a young engineering student at the Munich Technische Hochschule attended, in 1875, a course of lectures on the elements of thermodynamics. Many had done so before him, and very many more were to do so afterwards. But this young student was different for, when the Carnot cycle was expounded as the unapproachable ideal he made a note to try to devise some sort of engine which would approximate much more closely to the Carnot ideal than any of the heat-engines then in use. The student's name was Rudolf Diesel.

Diesel realised that for practical purposes the deciding factor was the fourth condition set out by Beau de Rochas: that the greatest possible compression should precede the start of the cycle. The engine he envisaged was to work in this manner (Figure 31): the first stroke of the piston drew air into the cylinder and the second stroke, the return, compressed the air to such an extent that the temperature rose 'adiabatically' far above the ignition point of the fuel. When the piston had reached the end of its travel and the pressure and temperature of the air were at their highest a fuel port opened and a jet of

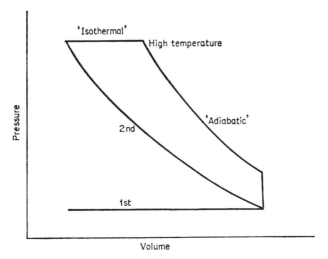

Figure 31.

coal dust (carried by a blast of compressed air) or a spray of oil was injected. Because of the high temperature of the air in the cylinder the fuel ignited. But the piston was now moving outwards and as the air was heated by the ignited fuel so the expansion, allowed and caused by the movement of the piston, prevented an increase in temperature; the heat developed by each particle of fuel being instantly absorbed by the 'adiabatic' cooling due to the expansion. Thus in the Diesel cycle the heating does not take place in one bang as it were, at constant volume, as in the Otto and Beau de Rochas cycle but at constant pressure and isothermally so that there is no rise in temperature of the working substance due to the burning of the fuel. In Carnot's language there is no useless flow of heat from a hot body (the burning fuel) to a 'cold' body (the highly compressed air).

Following the isothermal expansion of the air at constant pressure, the fuel supply is turned off and the expansion of the air continues 'adiabatically'; the four-stroke cycle is completed by the expulsion of the burned fuel and the recharging of the cylinder with air. Although it involves some repetition it is worth quoting Diesel's own description of this cycle:

> 1. Production of the highest temperature of the cycle (temperature of combustion) not by and during combustion, but before and independently of it, entirely by compression of ordinary air.
>
> 2. Gradual introduction of finely divided coal into the mass of highly compressed and therefore highly heated air, during part of the return stroke of the piston. The combustible is added in such a way that no increase in temperature of the gases, consequent upon the process of combustion, takes place and therefore the compression curve approximates closely to an isothermal. After ignition, combustion should . . . be regulated by an external arrangement, maintaining the right proportion between the pressure, volumes and temperatures.
>
> 3. Correct choice of the proper weight of air in proportion to the thermal value of the fuel according to . . . (a given formula); the limit of the compression temperature T_1 (which is equal to that of combustion) being previously determined so that it is possible to work and lubricate the engine without artificially cooling the cylinder walls.

Diesel was soon forced to give up coal dust as a fuel: the difficulties of lubrication proved too great. In 1897 the first oil Diesel engine was built at the M. A. N. works at Augsburg and thereafter the Diesel engine has proved increasingly successful as an extremely economical and versatile source of power.

The Diesel engine is a very solid machine, designed to stand high pressures and built to very exacting standards. Diesel did not invent—and never claimed to have invented—the principle of igniting the fuel by compressing the working substance, the air. That principle had been invented before his time. The essence of his invention was that the air was compressed and therefore heated far above the ignition temperature of the fuel after which, and only after which, the fuel was injected so that isothermal expansion could take place. For with isothermal expansion all the heat absorbed is converted into external work.

It is important to make this point for Diesel has sometimes been denied the credit for his invention.* No doubt other and quite capable engineers were working on engines which bore some resemblance in form to the Diesel engine. But unless and until they can be shown to have had the complete grasp of the fundamental principles which Diesel obviously possessed he, and not they, must be regarded as the true inventor of the engine that bears his name. For our purposes it is enough to say that at least a man whose scientific insights and engineering standards were comparable with those of Watt had applied his talents to the development of the air-engine. His appreciation of the possibilities of his engine was certainly accurate for in 1894 he wrote:

> We consider the new motor especially applicable to railways, to replace ordinary steam locomotives, not only on account of its great economy of fuel, but because there is no boiler. In fact the day may possibly come when it may completely change the present system of steam locomotion on existing lines of rails.

Diesel died in mysterious and tragic circumstances when he was lost overboard from the Harwich–Hook of Holland boat in 1913.

So far we have considered only gas- and heavy-oil engines. The invention of the carburettor, however, also enabled engines to be developed that used very light oils such as petroleum. In 1885 Gottlieb Daimler, an Austrian, obtained an English patent for such an engine working on the Beau de Rochas cycle and which had many of the characteristic features of the modern automobile engine. This high-speed engine could be made much smaller and lighter without sacrifice of power, and Daimler was able to use it to power a bicycle, a small boat and, in 1887, a carriage. Benz and Maybach were other pioneers in this field of the lightweight high-speed petrol-engine. The

---

* Arthur Evans, in his *History of the Oil Engine*, credits the Diesel engine to Emil Capataine who, he says, sold his patent to Diesel. Evans made no attempt to prove his case and gave no evidence that Capataine was working on the very advanced lines of Diesel.

motor car had, in principle at any rate, been invented by Daimler; the aeroplane was at last in sight as an engine within Cayley's specification became increasingly feasible.

A wide range of power requirements which could not possibly be met by steam-engines could, by the end of the nineteenth century, be satisfied by the newly developed air-engines. Even on its own ground, the provision of substantial power for factories, mills, railways and ships, the steam-engine was about to be challenged by the Diesel engine. If this was not enough yet another challenger emerged when, at last, the electric power industry got under way in the closing decades of the century.

In principle the generator and the motor had existed since 1831. In economic terms, however, they remained hardly more than toys for the steam-engine and the gas- and oil-engine could meet all requirements. The only practical uses for electricity were those things for which it was pre-eminently suited: electroplating and telegraphy. The development of transoceanic cables, following the successful laying, after a number of fruitless attempts, of the great Atlantic cable in 1865, meant that in a sense the world contracted very rapidly indeed. It now became a matter of seconds, not weeks or months, to transmit messages and news from continent to continent. In social, political and economic affairs this was immensely important. Technologically its main effect seems to have been the establishment of specialised firms for the manufacture of electrical apparatus—meters, insulators, coils and cable—and the emergence of a new class of professional technologists, the telegraph engineers.

The great breakthrough, however, occurred with the invention by Joseph Swan and T. A. Edison of the incandescent electric light bulb. A number of methods of using the electric arc as a source of light had been tried during the nineteenth century. The arc lamp and Jablockhoff candles had had some success but they could be used only for lighting streets, public squares or large yards. The invention of the small incandescent bulb with its agreeable light and absence of noise, odour or danger of any sort changed the picture completely. It could be used in ordinary homes and in public buildings. There was an immediate demand for electric lighting and, accordingly, for power stations and transmission lines to supply the electricity. In this, the early phase of the industry in Britain, a leading part was taken by an able young man of Italian descent, Sebastian de Ferranti. He saw the need for big, centralised power stations and put up the first one of this kind at Deptford; he developed underground transmission lines working at a high voltage and he settled on the 50 cycles, alternating current system. Later Ferranti moved north to Manchester and founded what was to become one of the largest electrical companies in the country.

In one sense electricity was not a competitor, at least not an immediate competitor, with steam power. Its first rival was the gas industry, in which field even the belated invention of the Welsbach gas mantle in place of the old fish-tail burner could not halt the triumphal progress of electric lighting. Electricity, in any case, had to be generated and this necessitated steam-engines or, in hilly countries, water turbines. The first power stations, such as the one at Deptford, employed large Corliss-type steam-engines coupled by rope drive to big alternators. In this respect the power stations were merely copying the power technology of the cotton mills which had already adopted rope drive and the Corliss steam-engine from America.

However, reciprocating engines, no matter how efficient, were unsuitable for electrical generation. This had been clear from the beginning for the problems of vibration were inseparable from the high-speed drive required by the alternators. It was to meet this challenge that (Sir) Charles Algernon Parsons, a younger son of the distinguished astronomer and President of the Royal Society, the Earl of Rosse,* developed a successful steam-turbine in 1884. In this machine steam is generated at a very high pressure and therefore at a very high temperature, after which it is directed, by a series of nozzles arranged in a circle, against the blades of a rotor. The blades are suitably curved so that the steam impinges smoothly and without shock, giving up its energy to the rotor (Figure 32). Thereafter the steam can pass, at lower pressures, through a second and a third rotor, progressively expanding and giving

---

* The Earl of Rosse installed a large telescope on his estate at Parsonstown, in Ireland (Eire) and used it to identify the spiral nebulae that are the individual stellar continents of modern cosmology.

up its energy in stages. Finally, the steam will be at the same pressure (and temperature) as the condenser, so that when it leaves the blades it will be at rest relative to the nozzles and the casing. In this way the maximum possible efficiency is attained and the whole of the energy of the steam is given up to the rotor. The turbine, we see, can be designed in accordance with the conditions for the maximum thermodynamic efficiency and it has the additional advantage, in terms of mechanics, of direct rotative, instead of reciprocating motion. It was, in fact, the long-sought-for solution of the problem of the direct rotative steam-engine, first envisaged by Watt in 1769.

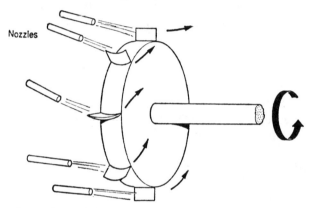

Nozzles

Figure 32.   *The steam turbine.*

The steam-turbine was invented when there was an appropriate demand for very high-speed drive, when engineering techniques were sufficiently advanced to enable the manufacture of such engines and, lastly, when there was a man with the necessary vision, drive and means to achieve it. The turbine was applicable not only to the generation of electric power, undreamed of in 1769, but also to manufactures and transport. At the first of these tasks, however, it has not been competitive with the more compact and hardly less efficient electric motor; and at the second its success has been limited to marine propulsion. No steam-turbine road vehicle has been developed and while a number of steam-turbine railway locomotives have been built they have invariably proved to be too complex and therefore too costly to maintain. Furthermore, the severe design limitations, imposed by railroad requirements, have made them marginally, if at all, more economical than the familiar reciprocating steam locomotive of the recent past.

THE ORIGINS OF MODERN COMMUNICATIONS

The rapid development of the electric telegraph from the 1840s onwards had accustomed men to the marvel of near-instantaneous communications across oceans and continents. Almost inevitably, it seems, the art was extended to include the transmission of speech when, in 1875, Alexander Graham Bell invented the first effective telephone. Sound waves in air cause a microphone to superimpose upon an electric current flowing through it, ripples or waves of the same frequency, or pitch, as the sound waves and of an amplitude or size proportional to the loudness of the sound. By passing this 'modulated' current through an earphone, consisting of a small electromagnet placed near a flexible metal diaphragm, the sound can be reproduced at the other end of the circuit, or 'line'.

The rise of telegraphy and telephony stimulated new industries and techniques for the design and manufacture of components and, at the same time, brought about the rise of a new class of professional men: the telegraph engineers. Naturally there were those who dreamed of even more advanced forms of telegraphy. As early as 1838 K. A. Steinheil of Munich had shown that one of the two wires used in overground telegraphy could be dispensed with by using 'earth return' to complete the circuit. A shovelful of earth may have a comparatively high resistance to the passage of electricity, but the countless shovelfuls that constitute *terra firma* when taken together offer a very small resistance between two points on the surface of the globe. Steinheil even looked forward to the day when the other wire might be abolished and a completely 'wireless' telegraphy established.

Apart from attempts to use the conducting powers of earth and water, a form of wireless telegraphy was already conceivable on the principle of Faraday's electromagnetic induction. If one circuit contains a telegraph key and an oscillator to generate a current of 'audio' frequency,* then the continuously changing magnetic field due to this current will induce an alternating current in another and quite separate circuit. If the second circuit contains an earphone then Morse code signals can be sent from the primary, or transmitting circuit to the secondary, or receiving circuit without the use of connecting wires. Among those who experimented with wireless telegraphy of this sort were Professor Trowbridge in the U.S.A. and Oliver Heaviside in Britain. On the other hand Thomas Alva Edison experimented with a form of wireless telegraphy based on the principle of electrostatic induction.

Any system of inductive wireless telegraphy would have required very sensitive detectors or earphones for the electrical energy, instead of being guided or canalised by wires from the transmitting to the receiving station, would be broadcast impartially in all directions and only a minute fraction of the total sent out would be available at any particular distant receiver. It was, in fact, computed that enormous transmitting and receiving circuits, as big as the British Isles or the State of New York, would be required to send detectable messages across the Atlantic ocean. But communications over shorter distances and to stations not easily reached by cable—such as lighthouses and lightships—were entirely feasible, both practically and economically.

The theoretical possibilities were there; so too were the vision and the practical incentives. But, as it happened, without another and most fundamental advance in science, as distinct from technology, modern communications—radio, radar, television—could not have been achieved. If we agree that thermodynamics was a gift from the power technologies to science and philosophy, the contemporaneous development of electromagnetic field theory was to prove no less important a gift, but in the opposite direction. We must therefore pause for a while to consider the scientific advance that made this possible.

Sir Isaac Newton had regarded the space between two bodies that attract each other as being of no significance. The earth's gravity holds the moon in orbit, but good Newtonians

---

* That is, an alternating current which, if applied to an earphone, produces sounds of a frequency audible to the human ear.

did not bother about the ultimate cause of gravity—what goes on between the earth and the moon—they were satisfied with the idea of 'action-at-a-distance'. However, with the gradual acceptance of the 'undulatory' theory of light from the beginning of the nineteenth century onwards, the idea that the intervening medium (whatever it might be) was important began to grow in men's minds. If light travelling from a bright source to the eye of the observer is supposed to consist of undulations, or very rapid vibrations, it is commonsense to ask what it is that is supposed to be vibrating. Sound waves are plainly rapid vibrations of air for we know that sound cannot travel through a vacuum, where there is no air. Breakers on a beach are inconceivable without a sea to move up and down. What, then, is it that moves up and down or vibrates when a ray of light travels outwards from its source? The answer that nineteenth-century science gave, with increasing conviction, was that the vibrations that caused the phenomena of light were vibrations of an elastic aether. The aether was supposed to pervade all space, otherwise we should be unable to see the sun and the stars, as well as our atmosphere. It produced no other effects save those of light and radiant heat. All that we knew about it was that it must be very rigid and very elastic in order that the ultra-rapid vibrations of light and heat can be transmitted by it.

On the other hand two separate developments suggested that there must be a close relationship between light and electro-magnetism. Michael Faraday had shown that a powerful electromagnet could produce a marked effect on a ray of light: a strong magnetic field deflected the 'plane of polarisation' of a suitably 'polarised' ray of light. Faraday realised that this effect indicated that light was, in some way, electromagnetic in nature. But from being convinced of this to being able to show how and in what way light was electromagnetic was an enormous step that not even Faraday, with all his genius, could take.

The second development was no less tantalising and even more unexpected. It had been easy enough to find a rational basis for standardising a unit of electric charge: it should be one that exerted a unit force on an identical charge placed one centimetre—a unit distance—away. Similarly, a unit of electrical current should be one that, flowing along a centimetre of wire bent into an arc of a circle of one centimetre radius, exerted a unit force on a unit magnetic pole placed at the centre of the circle. But a unit current is, by definition, one that carries a unit charge past a point on the wire every second. So we have two sets of standard units for measuring electric charge, electric current and, indeed, all other electrical quantities: 'electrostatic' units and 'electromagnetic' units, both expressed in terms of mechanical force. In 1856-7 two German scientists, W. E. Weber and R. H. A. Kohlrausch carefully measured the same electric charge first in one set of units and then in the other. Electrostatic units are very much smaller than electromagnetic units and they found that if they divided the size of the charge in electrostatic units by its size in electromagnetic units they got a number that was nearly equal to the value of the velocity of light as recently measured by the French scientist Fizeau.*

The man who put together the pieces of this natural jigsaw was James Clerk Maxwell (1831-1879). He began by trying to picture Michael Faraday's magnetic field in the neighbourhood of a wire carrying a current as essentially dynamic for the lines of force vanish immediately the current stops flowing. Maxwell suggested that when a current, consisting of individual charges, starts to flow along a wire it causes a series of vortices or whirlpools to spin round in the surrounding aether. The axes of these vortices correspond to the lines of force and the direction in which they rotate—clockwise or counterclockwise —determines the direction of the lines of force. The vortices exert centrifugal force as they spin and so press outwards on each other; at the same time they tend to contract so that there is always a tension along their axes; along, that is, the lines of force (see Figure 33).

Once the vortices next to the wire have been set spinning their motion must somehow be transmitted to the next series and so on. But this must be done in such a way that all the

---

* Electromagnetic measures involve flow, or motion; electrostatic measures do not. When the dimensions in which the units are measured are taken into account it is found that the ratio of the two reduce to a length divided by a time. That is, the ratio has the dimensions of a velocity and the magnitude of the velocity of light.

vortices spin in the same direction. To make this happen Maxwell supposed that each vortex was separated from its neighbours by a number of small particles that acted like 'idle wheels'; that is, they transmitted the motion from vortex to vortex in such a way that all the vortices rotated in the same direction.

So far, so good: but Maxwell's aether has to be elastic. He therefore allowed the idle wheels, or small particles, a limited degree of freedom of bodily motion. When a current starts to

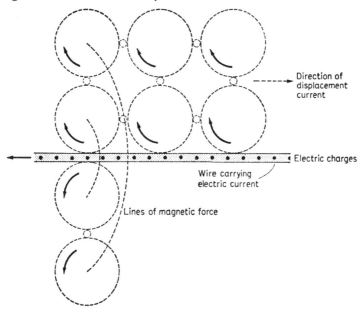

Figure 33.

flow in a wire and the nearest vortices start to turn the particles round their edges start to move orbitally, like epicyclic gears. Those furthest from the wire naturally tend to move in the opposite direction to the starting current, but before long the restraining forces of the elastic aether stop them and they must start to rotate, transmitting the spin of the innermost vortices to the next outer layer.

If we now regard these idle wheels as elementary charges of electricity we can understand the nature and process of electromagnetic induction. The rotation of the vortices will cause a momentary current to flow in a wire placed near the one in which the current has just started. The secondary current will be in the opposite direction to the primary current and it will soon cease for the wire will not be a perfect conductor. When the primary current is switched off the continued rotation of the outer series of vortices will cause a momentary current to flow in the secondary wire as the inner series of vortices will have stopped with the primary current. As Maxwell put it:

> It appears therefore that the phenomena of induced currents are part of the process of communicating the rotatory motion of the vortices from one part of the field to another.

In this way Maxwell constructed a model of the electromagnetic field that accounted for the observed phenomena of induction. It was also agreeably Victorian; like a large and busy workshop with pulleys, gears and line-shafting all industriously turning. It was not only Victorian: it was very English* and in the apparent crudity of conception it shocked gently and philosophically nurtured foreign scientists. But Maxwell was not being dogmatic; he was developing his very profound thoughts on the role of the medium or aether:

> The conception of a particle having its motion connected with that of a vortex by perfect rolling contact may appear somewhat awkward. I do not bring it forward as a mode of connection existing in nature or even as that which I would willingly assent to as an electrical hypothesis. It is, however, a mode of connection which is mechanically conceivable and easily investigated. . . . I venture to say that anyone who understands the provisional and temporary character of this hypothesis will find himself helped rather than hindered by it in his search after the true interpretation of the phenomena.

Maxwell then went on to extend his exploratory hypothesis to the electrostatic case. If an electric tension, or electromotive force, is applied to the aether, or to any non-conductor, we may expect that the small particles, or idle wheels, will be displaced from their normal rest positions to the extent that the elastic

---

* Maxwell was, of course, a Scotsman; but most European scientists seemed to have been incapable of distinguishing between Scotsmen and Englishmen.

restoring force permits; and that they will remain so displaced for as long as the tension is applied. When they are actually being displaced—for that moment when the tension is being switched on—the drift of particles constitutes the beginning of an electric current; but it is a current that stops as soon as they reach the limit allowed them by the restoring force. This short-lived 'displacement' current (as Maxwell calls it) must bring about a transient magnetic field that will vanish as soon as the current stops. In its turn, the changing and temporary magnetic field must—as Faraday had shown in 1831—bring about a transient electric tension or electromotive force that would cause an induced current to flow in a conducting circuit, or a brief displacement current to 'flow' in a non-conductor or in the aether. Thus a short-lived disturbance will spread outwards through the aether from a source which can be either an electric tension being applied or an electric current starting to flow. It will be an electromagnetic disturbance, the electric field generating the magnetic field and *vice-versa*.

In his seminal paper of 1865, *A Dynamical Theory of the Electromagnetic Field*, Maxwell dispensed with the vortices and idle wheels but kept the basic and sovereign idea of the displacement current. He recapitulated the simple equations that expressed the way in which an electric current (which could be solely a displacement current) could give rise to a magnetic field and the way in which a changing magnetic field could give rise to an electric tension. When these simultaneous equations were simplified by leaving out the source current and considering only the displacement currents, he found that the resulting electromagnetic disturbance was propagated with a velocity that was the same as the ratio of the two sets of units and was therefore equal to the velocity of light. Light must therefore be an electromagnetic disturbance, characterised by lines of electric and magnetic force at right angles to each and to the direction in which the light is travelling.

Maxwell's views may be summed up very briefly: only in a steady state can a magnetic field exist without causing an electric field and *vice-versa*. When one is changing, it automatically brings the other into being for as long as the change continues (Figure 34). Always these mutually generating fields must be at right angles to each other (as his original model suggested) and they must both travel with exactly the same velocity, which is equal to that of light.

As far as Maxwell was concerned an all-important synthesis had been achieved, relating light and electromagnetism, and a general theory of the electromagnetic field had been established. And there the matter ended. It was, we must assume, of theoretical interest to him that an electromagnetic pulse is

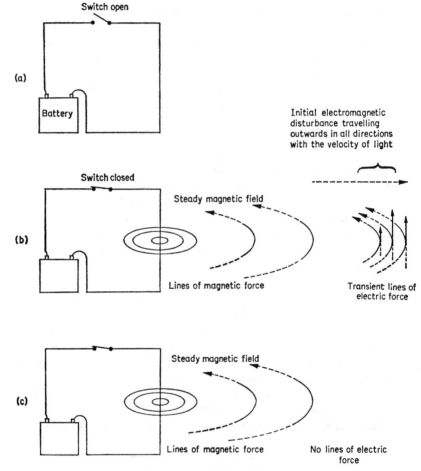

Figure 34.   (*a*) *Just before the action commences.*
                  (*b*) *Immediately after the current has started to flow in the circuit.*
                  (*c*) *When the steady-state magnetic field has been established.*

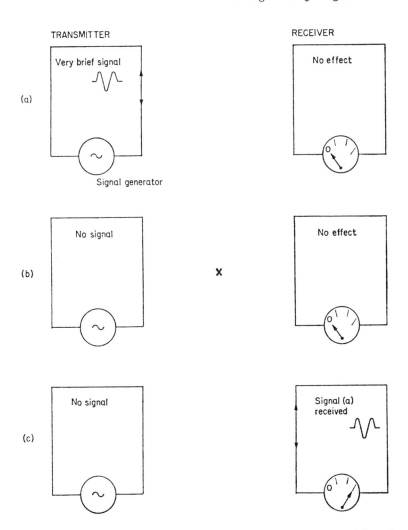

Figure 35. *If there is an appreciable time interval between stages* (a) *and*
(c) *there must be some activity in the space between the trans-*
mitter and receiver in stage (b). *This activity, which cor-*
responds to Maxwell's mutually generating electric and
magnetic fields, is represented by X. *If there is no such*
activity stage (b) *cannot occur and* (a) *must be accompanied*
without any time-lag, by (c). *This corresponds to the old,*
or classical theory.

G

transmitted when an electric current is switched on or when an electrostatic tension is established. This detail completed and generalised the theory of electric and magnetic fields sketched out by Faraday. But the actual detection of such very brief transients must, if he gave the problem any thought, have seemed an impossibly difficult experimental task. Still less could he have been expected to foresee any practical or commercial use for such radiations. Some men of science, among them G. F. Fitzgerald, O. J. Lodge and H. A. Lorentz, did speculate on the possibilities of detecting Maxwell-type electromagnetic disturbances but they found the practical difficulties insurmountable. As for Maxwell, he died prematurely in 1879.

Imposing though Maxwell's theory was it remained speculative and unproven for more than twenty years. It was, men argued, by no means necessarily correct; other theories, consistent with Newton's principles, might be possible and after all the electromagnetic nature of light had been indicated, if not demonstrated by the experiments of Michael Faraday and by the measurements carried out by Weber and Kohlrausch. Maxwell's theory, therefore, represented the wisdom of hindsight based on dubious theorising and some odd ideas about the nature of the aether. An acid test for it would be to show, experimentally, that a changing electric field generates a transient magnetic field and *vice-versa*. Accordingly, in 1879, the Berlin Academy of Sciences offered a prize to the experimentalist who could do this. The challenge was taken up by Heinrich Hertz (1857–1894), a pupil of the celebrated Hermann von Helmholtz.

Hertz realised that the problem would be solved and Maxwell's theory confirmed if he could show that electromagnetic waves, generated by a changing or oscillating electric current, travelled through space *with the same velocity as light*. According to the old, pre-Maxwell, theory the communication of inductive effects must be instantaneous and the velocity accordingly infinite. The point was put with great clarity by Henri Poincaré, the distinguished French mathematician (see Figure 35):

> . . . *according to the old theory the propagation of inductive effects should be instantaneous.** If, indeed, there be no displacement

* His italics.

currents, and consequently nothing, electrically speaking, in the dielectric that separates the inducing circuit from that in which the effects are induced, it must be admitted that the induced effect in the secondary circuit takes place at the same instant as the inducing cause in the primary; otherwise in the interval, if there were one, the effect is not yet produced in the secondary; and as there is nothing in the dielectric that separates the two circuits, there is nothing anywhere. Thus the instantaneous propagation of induction is a conclusion that the old theory cannot escape.

Here, then, was an *experimentum crucis*, a decisive experiment on the outcome of which (if it could be performed) Maxwell's theory would stand or fall. Let us, therefore, briefly enumerate the experimental tools that were available to Hertz. In 1850 Fizeau and Gounelle and, in 1875, Werner Siemens had shown that electricity travels along a wire with a velocity approximately equal to that of light. At about the same time Kirchhoff had deduced, using the old theory, that electricity should travel along a wire with a velocity that increases as the resistance is diminished until, when the wire has no resistance at all the velocity becomes exactly equal to that of light. On the other hand in the case of an alternating current the more rapid the oscillations—that is, the higher the frequency—the less important the resistance becomes and the more closely the velocity of the electrical oscillations approximates to that of light. Exactly the same inferences can be drawn from Maxwell's theory.

The problem therefore was to find a way of generating high frequency electrical oscillations and to compare the velocity of such oscillations travelling along a wire with that of the electromagnetic waves that, according to Maxwell, the oscillator should radiate into space. As early as 1842 Joseph Henry had called attention to the oscillatory discharge of a Leyden jar, or condenser: under certain conditions the electric charge on one plate of a condenser will surge through the connecting circuit and 'overshoot', piling up on the other plate until it starts to flow back, overshooting once again and piling up on the first plate. This oscillatory process continues, although with diminishing amplitude as resistance takes its toll, until at last the condenser is entirely discharged. In 1853 Kelvin provided a mathematical explanation of oscillatory discharges and showed how the frequency of the oscillations depended on such factors as the 'capacity' of the condenser. Finally, in 1858 Feddersen actually detected oscillatory currents using a spark gap, a rapidly rotating mirror and a camera. All that was needed in fact was an induction coil to produce high voltages, two pieces of wire with a short gap between them for a spark to jump across and a suitable detector to indicate the electromagnetic waves in space and the oscillations along a wire. The terminals of the induction coil are connected to the two pieces of wire and when a spark takes place an electric charge surges from one piece of wire to the other with a frequency that depends on the 'capacity' between the two pieces of wire. This capacity can be made very small and so the frequency can be made very high indeed.

In 1886 Hertz invented a simple detector for high frequency oscillations. It consisted of an adjustable loop of wire with a small spark gap in it. If this detector was brought near an oscillating circuit sparks could be obtained at the gap when the loop was held in the correct position to obtain the maximum inductive effect. This detector could be varied in size and thus 'tuned' to the frequency of the oscillator so that the maximum response was obtained; it was therefore called a 'resonator'.

With very simple apparatus (Figure 36) consisting of a spark-gap oscillator, a length of wire and a tuned resonator Hertz carried out a brilliant series of experiments (1888–1889). After some difficulties he succeeded in showing that the wave length of oscillations transmitted along an open wire was equal to that of electromagnetic waves in open space. As the same oscillator working at the same frequency was used in both sets of measurements it followed that the velocity of the waves along the wire was equal to that of the waves in open space. This common velocity was shown, by both the above arguments and by Hertz's measurements, to be the same as that of light. The *experimentum crucis* had therefore decided firmly in favour of Maxwell's theory; and the reality of Maxwell-type electromagnetic waves had been established.

Thereafter Hertz went on to show that these waves had the same physical properties as the vibrations that caused the sensation and phenomena of light. They could be reflected and refracted; they were 'normally polarised' and could be made to display the characteristic phenomenon of 'interference'. Hertz had enormously extended the spectrum, from visible light through radiant heat to the new Maxwell-type waves whose length, from crest to crest, was measured in metres, rather than in fractions of a millimetre. Significantly, too, Hertz showed, by experiment and theory, that the 'field strength' of electromagnetic waves diminished much less rapidly with the distance than did the field strength of Faraday-type induction (see pp. 147–8 above).

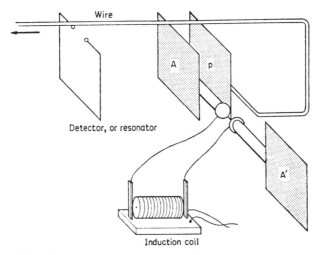

Figure 36.   *Hertz's apparatus as used in his early, critical experiments.*

On 1st January 1894 Hertz died, tragically young. Five months later, to the day, Oliver Lodge delivered a commemorative lecture at the Royal Institution in London, outlining Hertz's achievements and discussing the latest techniques for detecting and studying electromagnetic or, as they came to be called, 'Hertzian waves'. The 'coherer', as Lodge called it, had been invented by Edouard Branly. It consisted of a tube filled with powdered metal and with a terminal at each end. Normally the resistance of a coherer was very great but if Hertzian waves impinged on it the resistance fell drastically, for the particles of metal seemed to stick together, or 'cohere'. Lodge therefore placed a coherer in the same circuit as a battery and an electric bell so that if it was exposed to a short burst, or train of Hertzian waves, the resistance of the coherer fell and the electric bell rang. Moreover, the coherer was placed on the same stand as the electric bell so that the vibrations of the latter broke up the coherence of the particles and thus reset the instrument for the arrival of the next signal.

Much valuable work was being done on Hertzian waves at this time but it seems to have been almost entirely in the domain of physics rather than electrical engineering and telegraphy. It was left to a wealthy young Italian, barely out of his 'teens, to convert these highly scientific developments into a new and revolutionary method of transmitting information. Guglielmo Marconi (1874–1937), had attended Augusto Righi's lectures at Bologna on Maxwell's theory and Hertz's experiments and he had read Lodge's Royal Institution lecture. He could, he said, hardly credit that the great men of science of the time had not already seen the practical implications of Hertzian waves; but, as Lodge later confessed, the remarkable fact was that they had not.

Marconi carried out his first experiments on his father's estate. He greatly improved the coherer, finding a mixture of 95 per cent nickel and 5 per cent silver the most suitable, and he developed the 'aerial' or 'antenna' in place of the two bits of wire, or 'dipole' that Hertz had used. He found, too, that he could greatly increase the range of reception by using high transmitting and receiving aerials. Convinced of the commercial possibilities of his inventions the young Marconi packed his bags and set off for England. In June 1896 he lodged his patent—the first wireless patent—and established the Marconi Company with an issued capital of £100,000. The issue was immediately subscribed and cable and telegraph shares took a nasty tumble on the stock exchange. As well they might, for five years later, in 1901, Marconi succeeded in establishing wireless communication across the Atlantic ocean.

Marconi was, of course, lucky: his mother was a wealthy Englishwoman and, no doubt, she opened many doors, official and commercial, in London. He never seems to have suffered from those long spells of frustration, failure and poverty that seem to have been the common lot of inventors; he was, on the contrary, successful from the very beginning. Nevertheless there can be no doubt that Marconi was the inventor of wireless. He alone had understood the immense practical possibilities of the Hertzian waves. He had come to England because that country was then the foremost maritime power in the world. More than half the world's shipping was registered in the United Kingdom; British shipowners should therefore be the first to appreciate the advantages of wireless telegraphy.* There are, in fact, close parallels between the achievements of Marconi on the one hand and Thomas Newcomen on the other. Both men established a revolutionary invention on the basis of the scientific work of their immediate predecessors; both men applied their inventions to those fields where the practical need was greatest—shipping and mining. And in both cases the applications of the original inventions soon spread far outside the limits of their first utilisation; and in doing so have transformed the conditions of civilised life.

Two other claimants to be considered the inventor of wireless telegraphy are the Englishman David Edward Hughes and the Russian A. S. Popov. But Hughes, who certainly demonstrated a very short range form of 'wireless telegraphy' in 1879–1880 gave no indication that he understood the nature of the 'electric waves' he experimented with; and when he was told by the informal committee of the Royal Society that investigated his claim that he was dealing with the well-known phenomenon of Faraday-type induction he lost heart and refused even to publish an account of his experiments though invited to do so. Hughes was not, therefore, a pioneer of wireless telegraphy. Similarly A. S. Popov, who had certainly read Lodge's lecture, invented a receiver of Hertzian waves but he used it to detect the electromagnetic radiations from lightning flashes. He therefore established an important new branch of science. But he does not seem to have invented a transmitter of Hertzian waves and nor does he seem to have appreciated their possible value for the communication of information. Popov

---

* What the early railway system was to the telegraph, ships were to wireless. It will be recalled that the first instance of the use of wireless to catch the attention of the general public was the arrest of Dr Crippen when his ship docked in Canada (1910). Two years later wireless played a vital role in the rescue of the survivors of the *Titanic* disaster.

cannot therefore be regarded as an inventor of wireless telegraphy, *per se.*

Marconi had soon discovered that for long distance communication, waves of much greater length than those used by Hertz were necessary. The length, from crest to crest, was measured in hundreds, even thousands of metres, whereas Hertz had used waves of only four or five metres' length. It was not long before Marconi noticed that reception tended to be much clearer by night than by day. And it was from observations of this sort that Heaviside in Britain and Kennelly in the U.S.A. inferred the existence of a 'reflecting layer', consisting of ionised gaseous atoms, some fifty miles above the surface of the earth. This 'ionosphere' in effect made wireless communications between Europe and America possible by 'guiding' the Hertzian waves round the curvature of the earth. Technology had begun to repay, in some measure, the debt that it owed to science.

THE NINETEENTH CENTURY—A RETROSPECT

The nineteenth century was a remarkably creative epoch as far as science and technology were concerned. At the beginning of the century Richard Trevithick was experimenting with the first steam railway locomotive; as the century ended Marconi was preparing to send the first radio signals across the Atlantic. In Trevithick's day Britain was the only industrialised country in the world; by Marconi's time several European countries together with the United States had industrialised themselves and Japan was taking important steps in that direction. Britain, in fact, had begun to slip behind in the technological race. It is worth discussing how and why this came about.

In 1854 the remarkably interesting engineering firm of Beyer, Peacock and Company was founded in Manchester. Carl Beyer and Richard Peacock were two of the five founders of the Institution of Mechanical Engineers and the firm that they established was one of the most energetic and progressive locomotive manufacturers in Britain. But the firm was also interesting—and significant—in that Carl Friedrich Beyer was a young German who had been taught at a *technische hochschule* and had come over to Britain to complete the practical side of his training as an engineer. He stayed on to become a wealthy and respected citizen of Manchester. Perhaps no less significant, if less impressive, was the fact that of the eight draughtsmen on the firm's books at the beginning no fewer than three were Germans, or had German names.

The 1851 exhibition had been a triumph for Britain. Of that there could be no doubt. But there had been disturbing signs that Continental Europe was taking steps to catch up with, and if possible surpass, Britain. Far-seeing men like Lyon Playfair started to call attention to the importance of industrial instruction and as a result of the Exhibition and of the lessons it taught, or was believed to have taught, the Science and Art Department was set up to encourage scientific and technical teaching in schools. Steps were taken to establish a college of science and technology in South Kensington; after a number of mutations this eventually became Imperial College.

Thereafter the history of technical education in England is one of increasingly urgent attempts to modernise on the basis of the old Mechanics' Institutes and the centralised systems of examinations run by the Science and Art Department and by the Royal Society of Arts. In 1869 a Parliamentary Select Committee studied scientific instruction and in 1872–1875 the Devonshire Commission made a very exhaustive examination of scientific and technical instruction, mainly in colleges and universities. In 1882–1884 the Royal Commission on Technical Instruction carried out a comprehensive, indeed world-wide, inquiry into technical education at all levels and in all countries, including the United States. As a result of their recommendations a system of technical education was set up which was based on evening courses given in technical colleges supported by local authorities. This system lasted until comparatively recently and traces of it are still apparent in the organisation of our local technical colleges. The Technical Instruction Commission was followed by other and more specialised ones dealing with universities and secondary schools. Paralleling these official actions was a strong voluntary movement for technical education which had the support of some of the most articulate engineers and scientific men of the day. This movement was instrumental in getting the wealthy London Guilds, or Livery

Companies, to make some of their funds available for technical education. In particular a number of specialised chairs were endowed at various provincial universities and the City and Guilds Engineering College, later a component of Imperial College, was founded.

All this action was taken in response to a situation that was explicit in the case of the firm of Beyer Peacock and which became increasingly explicit in other firms and industries as time went on. Britain, in a few words, was rapidly and obviously losing her lead.

The history of the heat-engine exemplifies the loss of leadership with sharp, indeed painful, clarity. Between 1712 and 1850 virtually every improvement in the heat-engine and every name associated with it—with the exception of a few writers like Prony, de Pambours and Carnot—was British. A visitor to the 1851 Exhibition could look at any one of the engines on display and know that every principle, every detail, every refinement was due to British genius. This would not have been chauvinism; it would have been undeniable, universally conceded fact. But after 1851 the situation changed decisively. Increasingly, improvements and new departures were due to the genius of Frenchmen, Germans and Americans. Beau de Rochas, Lenoir, Otto, Maybach, Daimler, Benz, Diesel; these were the great names in the development of the gas-, oil- and petrol-engines. Admittedly one can find British names too, but they were hardly in the international league. Parallel with the British failure to hold the lead, much less the monopoly, on the engineering side went a failure on the theoretical or scientific side too. Whereas the science of thermodynamics had owed a great deal to British engineers and scientists such as Black and Watt, Davies Gilbert, Woolf, Southern, Dalton, Herapath, Joule, Kelvin, Rankine and Maxwell; after 1851, and certainly after Maxwell, the main centres for the study of thermodynamics were European universities. So able an engineer-scientist as Osborne Reynolds made very little of thermodynamics; the new leaders were Helmholtz, Clausius, Boltzmann, Willard Gibbs, Wien, Stefan, Nernst and Planck.

The loss of technical leadership seems, as we have implied, to have been general, across the board as it were. Parsons' invention of the steam turbine might be thought to be an exception but it must always be remembered that Parsons was exceptionally favoured by his scientific background and by the financial independence he enjoyed. And even with these advantages he had a hard struggle with an obstinate British Admiralty before his design for a marine turbine was accepted by them. On the other side we have the fact that the best known steam-engine of the latter part of the nineteenth century was the Corliss engine, which had been a sensation at the 1876 Philadelphia Exhibition. This engine, with its characteristic cylindrical, snap-action valve mechanism, was soon at work in England. And we should not forget that the ultimate refinement of the reciprocating steam-engine, the uniflow engine, was perfected by Johann Stumpf at the Charlottenberg Technical University in 1908.*

One of the best-known examples of loss of leadership by the British was the synthetic dyestuffs industry. Here was a new field, based on the rise of organic chemistry in the early nineteenth century. As it happens the first work in the field had been by a young Englishman, William Henry Perkin, who in 1856 was able, thanks to a judicious blend of foresight, chemical ability and entrepreneurial skills, to launch a new synthetic dyestuff, the so-called mauveine, one of the first of its type, on the world market. The basic raw material was coal-tar which was of course very readily available in England, the main market was the great and still rapidly growing British textile industry and there was enough capital and willingness to innovate to ensure success in the first two decades of Perkins' new industry. But by 1900 the industry had almost entirely emigrated to Germany and the few firms left in Britain were small and unimportant compared with such giants as Badische Anilin und Soda Fabrik or Baeyer and Company.

Even in the new electrical industry, to which Britain had contributed so much through the geniuses of Wheatstone,

* The last stationary steam engine to be installed in a cotton mill was erected in 1926. Many of these later engines incorporated the Corliss valve mechanism and some worked on the uniflow principle. An engine which includes the latter and which was exhibited at the Wembley Empire Exhibition of 1924 as an example of British engineering skill has recently been secured for the Manchester Science Museum. It is interesting that the two ultimate refinements of mill engines should both have been foreign in origin!

Faraday, Kelvin, Maxwell, Varley, Fleeming Jenkin and several others, a great deal of the innovation was now coming from abroad; from organisations such as Siemens Halske, A.E.G. and the General Electric Company of New York. In cable manufacture, the electrification of railways, street cars and tramways, and electric lighting Britain was outpaced. While in the newest industry of all—motor car construction— the lead had demonstrably been taken by France, Germany and America. The first powered flight, in 1905, was to be by two American pioneers Orville and Wilbur Wright. Over the same period a whole host of ingenious inventions, many of them 'empirical', were coming from America: in the latter part of the nineteenth century came innovations like the zip fastener, the sewing machine, barbed wire, the phonograph (gramophone), the Westinghouse continuous air brake, the sleeping car and business machines, together with detailed improvements beyond computation.

It is not excessive to say that as far as Great Britain was concerned the closing decades of the nineteenth century and the opening decade of the twentieth do not really merit that golden glow of power, assurance, progress and world authority that present nostalgia for a more secure and perhaps more complacent age too readily discerns. In brief, it seems to have been a time of sustained and cumulative failure. And this was increasingly recognised by those who were best qualified to judge the nation's technological and scientific progress. There were in the Cavendish laboratory at Cambridge and in Manchester University, two first-rate centres for scientific research in the field of physics; in Germany at the same time there were about half a dozen centres of equal importance in physics and at least as many again concerned with chemistry. Virtually every professor of chemistry in a British university before 1914 held a German doctorate.

What conclusions can we draw from this sad tale and what explanations, if any, can we put forward? In the first place it is surely obvious that no single explanation will cover all the facts: that no single factor which might be important in the case of one particular branch of technology or industry will necessarily also be important in the other branches. Nor should we forget that as nation after nation took hold of the idea of industrialisation so inevitably Britain would have to lose her predominance which she gained by being first in the field. When giants wake up, they presumably do so refreshed. The more perceptive Victorians understood this and saw clearly enough that no action on Britain's part could stop it. Indeed they were generous and enlightened enough to welcome the prospect. Nevertheless there are degrees of failure: or rather more precisely there is always a point at which inability to be first, or second or third—which is excusable—turns into complete failure—which is not. In some cases the British failure seems so complete that some explanations are demanded. Just as we had to put forward some tentative reasons to explain why France with all the high genius of her technology and science during the revolutionary period was unable to overtake, or even catch up with Britain, so now we must try to find reasons why Britain, with all the immense advantages of industrialisation, wealth, and the kind of know-how of her industrial population that so impressed observers like Andrew Ure, ultimately lost her place. In putting forward or recapitulating certain explanations we shall emphasise those that bear on science and technology. For the failure was not necessarily, at this time, an economic failure; it was a technological and scientific failure. And the failure to be first, or even among the first, in the new industries founded on technology and science, was not a happy omen for the future.

A redeeming feature was that Britain was still a very open society. The authority of the state was limited; the police enjoyed no more rights than did other citizens; there was no conscription and, in law, no discrimination between one creed, race or political philosophy and another. A man could enjoy the profits of his labour in peace and security. England, with her markets, her capital and her toleration was, therefore, a strong magnet for able Europeans who found the atmosphere at home a little restrictive. Accordingly Britain benefited in the later nineteenth century from an inflow of technological talent just as she had profited in the past. The observer is left with the feeling that had it not been for the efforts of Italians like Ferranti and Marconi, Germans and Swiss like C. F. Beyer, Sir William Siemens, Hans Renold, Heinrich Simon, Ivan Levinstein, Ludwig Mond and many others, there might have been virtually no innovation at all in Britain. Apart, that is, from the innovations due to those foreign enterprises that had established subsidiary manufacturing companies in England: firms such as the Mannesmann Brothers, the British United Shoe Machinery Makers, the British Westinghouse Company, the British Thompson-Houston Company and, later, the Ford Motor Company. Such dependence on imported skills was surely a clear indication of native technological inadequacy?

The most immediate contrast between Germany and Britain was the excellence through the greater part of the nineteenth century of German university and technical college education compared with what was available in Britain. From the start Germany had more universities and they were, in contrast with British universities, subsidised; indeed generously subsidised. Research ability was the touchstone of a German university career and no fields of learning were neglected. The efficient German educational system provided a complete range of technological personnel from senior engineers or research scientists with doctorates through to commercial clerks and travelling salesmen. Reuleaux had criticised the German exhibits at the Philadelphia Exhibition of 1876; but his words were heeded and the flexible education system made it comparatively easy to train the men who could remedy defects as they became apparent. In Britain with its voluntaristic education system, with its emphasis on night schools, this was much less easily done.

Professor Habbakuk, in a work of great penetration, has pointed to certain important differences between Britain and America during the nineteenth century that favoured the development of American technology. It is impossible to do justice in a few words to an argument that is expounded with great clarity and subtlety and that occupies the greater part of a book. But briefly the suggestion is that in America during the nineteenth century land was cheap and labour, especially skilled labour, expensive. In Britain the opposite conditions prevailed: land was expensive and labour was cheap. Accordingly there was in America a strong incentive to invent and apply labour saving machinery. Hence the characteristically ingenious machines of the period 1850–1914, and more recently things like the combine harvester and the fork-lift truck: the list can be extended almost indefinitely. Once a technology is established on this basis, it can grow by virtue of the 'chain effect'. Add a certain gift of God—great natural resources— and a sensible constitution designed to encourage the energetic and creative and one has a good recipe for technological and industrial growth.

Summarising the situation at the dawn of the twentieth century, it would seem likely that Britain had been overtaken as regards the science-based technologies by Germany with her excellent university and technical college system. There were also such intangible factors as the growing confidence and assertiveness that came with national unification after such a long time in the wilderness of political fragmentation; a unification that set the seal on the more practical economic benefits of the Zollverein. At the same time Britain was outpaced in empirical technology by America. The turning point in this case can be dated with some precision as the year 1865. It was during and through the great Civil War that American industrialisation was given an immense impetus and it was at about this time that the first industrial research laboratory was established in Germany.

While Britain's friends and rivals were busy overtaking her, the British indulged in what, in retrospect, was one of the least valuable national exercises: the frantic, at times hysterical, pursuit of an Empire that was increasingly meaningless. The nation that had given the world Watt and Arkwright, Trevithick, Stephenson, Maudslay, Whitworth, Dalton, Faraday, Joule, Kelvin and Maxwell now sought prestige in planting union flags over miles of jungle, desert and swamp; and glory in such names as those of Rhodes, Jameson and Milner.

In 1914 Britain appeared to be a country whose major industries were still cotton and wool textiles; coal mining and iron and steel: the staples of the industrial revolution. It was a country too in which new industries and technologies were certainly represented but were rather too often subsidiaries of American or German parent firms, or else were started by naturalised Englishmen.

Chapter 6

The Twentieth Century

A history of science fiction, when one comes to be written, should prove very instructive. It would of course be co-extensive with the practice of science and would therefore begin with seventeenth-century writers like Samuel Butler and other satirists of the Royal Society. It would refer to the works of Dean Swift and to such later period pieces as Mary Shelley's *Frankenstein*. But from the technological and sociological points of view the greatest interest would surely attach to the later nineteenth-century writers: to the superb Jules Verne* and the perceptive H. G. Wells. For these writers undeniably influenced their contemporaries and successors, and they offer us valuable insights into the expectations which informed and imaginative men had of the technology of their times.

We have already suggested that mid-nineteenth century thinkers had fairly strong reasons for supposing that technology was (or had become) basically evolutionary and that little more than piece-meal improvements could be looked for. For this attitude they would have had strong grounds in the contemporary philosophy of science. It would be unwise to assume that all, or even most, Victorian scientists subscribed to a simple mechanical, or Cartesian–dualistic, view of nature. But there can be little doubt that most would, after 1865, have regarded the physical, biological and perhaps even the sociological laws as established and immutable. Since the resources of the world were known in nature, if not in extent, there were evident limits to what man could expect from his technology. *Per contra*, it would be a very bold, if not foolish man, who today tried to set any bounds to the power of technology. Most of us, if asked for a straight answer,

---

* It is well worth reminding ourselves that Jules Verne identified the site for the first moon shot as being in mid-Florida, not far from Cape Kennedy (*De la terre à la lune*, Paris 1865).

would probably agree that the prospects for technology are limitless.

There are a number of possible explanations why this happened. In the first place there were the writings of the science fiction novelists who, when they found their conventional science dull, professional, competent and uninspiring, tried to imagine circumstances which broke the conventional restrictions. The worlds they imagined in which flying machines, space travel, even time travel had been achieved together with the pictures they drew of civilisations of the future may well have helped to plant in men's minds some idea of the infinite powers of technology.

Confirmation of the new philosophy, if one can call it that, came from the advance of technology itself. The motor car—the 'horseless carriage'—and the aeroplane—the 'flying machine'—which had been dreamed of for so long had at last been achieved; wireless telegraphy and telephony, which would surely have been inconceivable as recently as 1851, were established inventions. The twentieth century seemed to strike a new note in the march of technology; the oil/gas-engine was the characteristic source of power, electricity was the new agent.

Lastly there were the breaches with nineteenth-century physical science that occurred at about the turn of the century. The first of the fundamental particles, the electron, was discovered and was shown to be common to all atoms of all elements. Dalton's atom therefore had a structure and could not be the ultimate unit of matter. (In fact the discovery of the electron marked a reversion to, or confirmation of, the classical atomic doctrine held by Newton and Boyle; but the age was not historically minded and the point was missed.) The development of sub-atomic physics coupled with the quantum theory and relativity meant that, once again, physical science was being re-cast as it had been several times since the seventeenth century. It followed that the more dogmatic assertions of the Victorians on the nature of man and his universe could now be rejected. Scientifically it was a permissive age and although many of the generalisations made about Victorian science were—and are—unfair, the effect was to open still wider the prospects for technology.

If we turn to contemporary speculation in order to gain some idea of men's expectations of the technology of their times we find that in their predictions of the future, or rather their extrapolations of contemporary technological trends as they interpreted them, writers often made shrewd prophecies. Following the inventions of telegraphy and telephony, television could readily be imagined; air travel by heavier-than-air machines (usually powered by steam-engines and therefore boasting handsome funnels) was confidently predicted long before the Wright brothers' first flight. Even the atomic bomb was, it is claimed, forecast by H. G. Wells not very long after the beginnings of sub-atomic physics.

It is, however, a truly remarkable fact that on the very brink of an economic-technological revolution unparalleled in history no one foresaw the *universal* motor car and all that it was soon to imply. This failure on the part of informed and perceptive men to grasp the significance of what was going on under their very noses must make us suspicious of all attempts to forecast technological developments even one or two years ahead, much less ten or twenty. Of course it would have been profoundly difficult to have foreseen the real importance of the motor car; it would have required omniscient knowledge of petroleum geology, of the economics of refining, of automobile engineering, of the capacities and tastes of the generality of men. But the difficulties of prediction do not in any way excuse the failure of prophecy. Our complaint is not directed against the science fiction writers but against those who would assert, as an academic exercise, that they can predict future technological trends. The astrologers, in effect, thought this too, although, before the days of Kepler and Galileo, there were some grounds for their beliefs.

If the automobile was ever to become anything more than a 'horse-less carriage', a vehicle to take wealthy people from their homes to the nearest railway station and back, or perhaps for a gentle spin along the promenade, then certain very important requirements had to be fulfilled. For one thing the road system, neglected in England since Roman times, undeveloped in America, satisfactory perhaps only in France, would have to be renovated and put into such a condition that cars could use it. Here, the path was smoothed for the automobile by a quite remarkable social and technological event that took place in the closing decades of the nineteenth century. This was the vogue of the bicycle. The 'penny-farthing', which appeared in the 1870s was popular with athletic and daring gentlemen but it was of limited appeal in other respects. In 1885 the safety bicycle was introduced by J. K. Starley, of Coventry, which had two small wheels but which could be propelled at a reasonable speed by means of pedals, chains and gearing. The safety bicycle allowed, indeed encouraged, women as well as men to take to the wheel and the bicycle boom began. The social and political consequences of the bicycle, especially as regards the emancipation of women, to which it probably contributed more than the efforts of the 'suffragettes' and the politicians, have never really been fairly assessed. Technologically its consequences were also important.

As a result of the bicycle boom efforts were immediately made by substantial numbers of people to improve the roads, to provide reasonable wayside facilities for the cyclist and, of course, to bring down the price of bicycles by mass production techniques. Dr Dunlop, an Irish doctor, invented the pneumatic tyre for use with bicycles and a variety of new techniques were developed for their production. The suspension wheel, with its echoes of Cayley and even Hewes, was introduced and light tubular steel construction standardised. Systems of braking and head and tail lamps were invented and developed. The point has been made very clearly by Professor John Rae, the historian of the American automobile. He quotes Hiram Maxim:

> It has been the habit to give the gasoline engine all the credit for bringing the automobile—in my opinion this is the wrong explanation. We have had the steam engine for over a century. We could have built steam vehicles in 1880, or indeed in 1870. But we did not. We waited until 1895.
>
> The reason why we did not build road vehicles before this, in my opinion, was because the bicycle had not yet come in numbers and had not directed men's minds to the possibilities of long distance travel over the ordinary highway. We thought the railroad was good enough. The bicycle created a new demand which was beyond the ability of the railroad to supply.

Then it came about that the bicycle could not satisfy the demand which it had created. A mechanically propelled vehicle was wanted instead of a foot propelled one, and we know now that the automobile was the answer.

For clarity and brevity combined with penetration this piece of technological history is very hard to beat. The bicycle having paved the way technologically, administratively— think of road maps and signposts—and socially, the automobile was able to come into its own as soon as the efficient internal combustion engines were available. As it happened oil production was soaring year by year to meet the demand for lighting and heating oils. The lighter, dangerous fractions were something of a problem for there was no use to which they could safely be put. The automobile provided that use.

The very first motor cars resembled horse carriages, minus the shafts; but the craft of carriage building could not compete with the new industry of bicycle making. Accordingly, in England the motor-car industry soon settled down in the home of the bicycle: in Coventry and Birmingham where light steel tubing was made, where components like gears and lamps were manufactured and where suspension wheels with pneumatic tyres were produced. Some of the firms that had grown up to manufacture bicycles turned to motor bicycles, to tricycles or three wheelers and then to motor cars. Perhaps the best known of British car-makers, the late Lord Nuffield, started off in the bicycle world.

On the other hand the development of the commercial road vehicle, the lorry or truck and the motor bus, went on in those places that were associated more with the steam-engine and heavy engineering. The steam traction engine implied the steam lorry, first with solid rubber tyres and then with pneumatic tyres. This led to the motor lorry in due course. Thus we have a polarisation of the industry which persists to the present day; cars in the Midlands, motor bicycles and cycles in the Midlands, lorries and heavy commercial vehicles in the north.

While the motor industry was getting under way the new sub-atomic physics was leading to a new form of technology that could hardly have been envisaged at that time without invoking a sixth sense. The discovery of the electron and the measurement of its distinctive properties: its mass and its electric charge involved the development and use of a specific instrument, the cathode ray tube. Cathode rays, so called, amount to no more than copious streams of electrons emitted from the negative terminal or 'cathode' fixed in a glass vacuum tube and sufficiently heated to emit thermal electrons. The electrons are projected down the tube by having a positive terminal, or anode, placed further down. By proper design the electrons could be made to overshoot the anode and impinge on the end of the tube where they would cause the glass to fluoresce with a characteristic green glow; better still, a special zinc sulphide screen would glow with a white-ish light exposed to a stream of electrons. This instrument has been progressively developed from the early experiments of Michael Faraday, on the passage of electricity through vacua, to the refined measuring devices used by J. J. Thomson, and later by Karl Ferdinand Braun, the engineer who devised the first modern cathode ray tube.

If the possibilities of the cathode ray tube were not immediately apparent, the utility of the thermionic emission of the newly discovered particles, the electrons, from heated metals was appreciated more readily. A glass tube, exhausted of air and fitted with a cathode in the form of a length of wire* surrounded by a concentric metal cylinder, or anode, constituted a device of great potential value for the wireless engineer. It would pass an electric current in one direction only since electrons, being negative in charge, can travel from a negative cathode to a positive anode but not in the reverse direction. This was the basis of the 'oscillation valve' invented and patented by J. Ambrose Fleming in 1904. In the November of that year Fleming wrote to Marconi outlining the use of such a 'valve' in wireless telegraphy.

The coherer, it will be remembered, worked by using Hertzian waves to switch on the current in a circuit containing an electric bell, or similar indicating device. The oscillation valve, or 'diode' as it was later called, worked on a different principle: it rectified the rapidly oscillating currents induced in aerials by Hertzian waves, thus replacing an alternating

* The cathode can be heated by means of the current from a small battery.

current by a direct one. Now sound waves of the same frequency as Hertzian waves would be far above the audible range. But if the incoming Hertzian waves can be modulated, like the current flowing through a telephone mouthpiece or

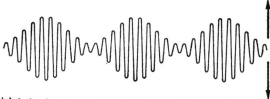

(a) A signal due to a modulated electromagnetic wave

(b) After rectification by a diode valve or a 'cat's whisker' crystal

(c) The resulting long—wave, or low-frequency, ripple or wave of current passing through earphones or loud-speaker to produce an audible noise

Figure 37.

microphone, then the rectified high frequency current will have an amplitude that varies relatively slowly, and passed through an earphone will produce audible sounds: Morse code or even human speech (Figure 37).

Fleming's diode was, however, almost immediately superseded by the 'cat's whisker' crystal detector (1906). It had been found that a crystal could act as a rectifier, passing current in one direction only, and as crystals were cheaper, smaller and more reliable than the early diodes they soon replaced the latter. The second stage in the development of wireless telegraphy and, a little later, telephony (or radio), was, therefore, that of the 'crystal set'. But the thermionic valve was not forgotten for a young American, Lee De Forest, was at that time working on parallel lines to Fleming. In 1907 De Forest invented a new, three electrode valve, or 'triode' as it came to be called.

If a thin mesh of wire be put between the cathode and anode of a diode valve, then a small change of voltage on the wire, or 'grid', could cause comparatively large changes in the current passing through the valve. The 'control grid' thus enabled the valve to act as an amplifier of signals so that a very small signal voltage on the grid could produce a very large one on the anode. During the years 1911–1914 the versatility of the new triode was increasingly realised. It could amplify weak signals, it could be used to generate high-frequency oscillations and it thus made possible the popular 'radio set' with loudspeaker in place of earphones; the instrument that, in the late 1920s, replaced the crystal set. The triode valve was followed by a whole series of multi-electrode valves: tetrodes, pentodes, hexodes and a wide variety of special purpose valves, developed for transmitting as well as receiving sets and later on for television and radar. The foundations of the new technology of *electronics* were therefore laid just before the first World War. As for the crystal set, it underwent a renaissance when, just after the Second World War, the transistor was introduced.

THE IMPACT OF WORLD WAR

In the light of subsequent events it is not easy to assess the consequences for technology of the First World War. The technologies associated with the internal combustion engine—the aeroplane and the motor car—received a boost and so did the different forms of wireless telegraphy. Generally speaking serious weaknesses were shown up in British technology, especially at the outbreak of war and at such episodes as Jutland. But it is quite outside the scope and intention of this work to recount the technological history of those four disastrous years. It is enough to say that as a result of war-time experience the Government of the country instituted the Department of Scientific and Industrial Research, now the Science Research Council, and undertook, in partnership with industry, the establishment of industrial Research Associations. The universities too were persuaded to institute such long overdue reforms as the establishment of the Ph.D. research degree, in direct imitation of what was now recognised as the successful German practice.

One post-war trend was unmistakable: the development of the technology of control systems and the march towards increasingly automatic production processes. Self-controlling machines which continually examined their own operations as it were and, if there was a deviation from the norm, auto-matically corrected it, had first appeared in the eighteenth century. The first ones were so idiosyncratic and specialised that no one could possibly have seen that there was anything in common between them. Thus one such device, which appeared in 1748, was designed to keep windmill sails always facing the wind. A small set of auxiliary sails, at right angles to the main ones, would only rotate if the latter were not facing the wind. When the auxiliary sails rotated they thereby caused, through suitable gearing, the top of the mill to turn until the main sails were facing the wind again, at which point the auxiliary sails stopped, having lost the wind. Another device was the governor, developed first for use with windmills but adapted (by Watt) for steam engines and, later still, applied to water power. But between these two devices there was little or no evident connection.

The post-war growth of the process industries—water supply, gas supply, petroleum refining—in which a continuous flow of material was subjected to various physical and chemical treatments, stimulated the invention and development of appropriate automatic control systems. If at any stage of the process something started going wrong, an appropriate device would measure the extent by which the product was deficient or in excess and feed back instructions to a controlling valve to correct the deviant factor: to raise or lower the temperature, to increase or reduce the pressure by varying the rate of flow, as the case demanded. These control systems were first of all pneumatically or hydraulically operated; later on electronic methods also were developed: the amplifying properties of the electron tube, or valve, being particularly useful for converting small measurement signals into large controlling ones.

It was not only in the process industries that strides towards automatic control were taken. As the demand for motor cars built up during the post-1918 period so the production side became more and more rationalised. The production of car components, cylinder blocks for example, necessitated a number of machining processes such as drilling, milling, boring and tapping and it was quite logical to place the necessary machine tools side by side so that the worker on one tool could easily pass the block on to his neighbour as soon as he had completed his part of the job. The next step was to put all the machine tools on to one base, to make them in effect one tool, although served by a number of workers. The final step was to dismiss all the workers but one and let the unified tool pass the block from stage to stage, automatically. This machine, the transfer machine as it came to be called, was first devised in 1924 when two were made for Morris Motors by the machine tools firms of Archdales and Asquiths. But they were before their time: the control systems were not sufficiently developed and in the case of Archdales at least the costs of developing and making the machine proved onerous in the extreme.

The automatic factory was on the way, even if the transfer machine was not finally developed until after the Second World War. It was perhaps in the light of all these developments of machine tools and automation, as it came to be called, that the late Charles Singer amended, unconsciously I suspect, Charles Babbage's words when, in 1951, he wrote:

> What is man? . . . Man makes things, but so do many animals; man shapes objects into tools, but so do a few animals; man alone makes tools with which to make other tools.

Manufacture by means of self-controlling tools which carefully measure the product against the acceptable standard and if found deviant automatically correct their own processes at the right stages is the final outcome, not yet consummated, of the new process of industrial production instituted nearly two hundred years ago by Richard Arkwright and James Watt. Rationalisation, mechanisation, control were the stages, and what was first most easily achievable in the continuous process industries, such as oil refining can now, we anticipate, be extended to flow production in such engineering industries as the manufacture of cars.

BRITAIN AND TECHNOLOGY

If the period 1860–1914 was one in which Britain lost her technological leadership and would apparently have lost her grip on technological progress had it not been for the welcome invasion of many talented foreigners, the period between the two World Wars and the post-war epoch from 1945 is a curious contrast. The reforms that were initiated in the period of decline slowly started to bear fruit—there were some signs that they were beginning to take effect before 1914—but it was a slow and difficult business to stop and then reverse a national trend especially when that trend had got such a hold that its effects had become apparent and alarming. The analogy of a malignant disease is perhaps not entirely inapt at this stage. Of this most recent period, then, it is reasonable to claim that British technology has been showing signs of autonomy and creativity again. But, and here was the rub, another analogy suggests uncomfortable parallels with France during the revolutionary years. State support for technology and science had certainly produced results, but the inventions and ideas were too often, it seems, not of immediate application.

The general exception to this rule was provided by the urgent military requirements of the country with the imperative demand that technology must be applied to defence. This was more than a matter of official policy and the directives of committees; it stimulated the inventive powers of individuals working on their own account.* Fundamental, if not of

* The development of the Schneider Trophy sea-planes and the Spitfires which they led to is too well known to be repeated here.

immediate application, were the efforts of a young R.A.F. officer named Whittle, who, in 1928, wrote a short thesis on the future development of the aircraft engine. High-speed inter-ceptor fighters (as they were called) were an essential feature of British defence against the bomber, which it was feared would 'always get through'. Whittle wrote:

> It seems that, as the turbine is the most efficient prime mover known, it is possible that it will be developed for aircraft especi-ally if some means of driving a turbine by petrol could be devised. A steam turbine is quite impracticable owing to the weight of the boilers, condensers, etc.
>
> A petrol-driven turbine would be more efficient than a steam turbine as there need be no loss of heat through the flues, all the exhaust going through the nozzles.
>
> The cycle for a petrol-driven turbine is . . . a constant pres-sure cycle. Air is compressed adiabatically . . . .

From this thesis sprang the jet engine which first powered an aeroplane in 1941. It is, of course, well known that Germany developed the jet engine independently and simultaneously so that the credit for this invention must plainly be shared. But it is interesting to note that Whittle's thoughts were along the lines established by the great pioneers of the heat-engine and that once again Britain was making a major contribution to power technology after so long in the wilderness.

The other side of the military defence coin was the detection of hostile aircraft. The destruction of such aircraft by inductive heating from transmitters was both financially and technolo-gically impracticable, no matter how attractive theoretically. But their detection and location in all weathers and at night was another matter. Once again a simultaneous invention was made: that of radar. This device worked on the principle of radio echoes: short pulsed radio signals together with the echoes were displayed on a cathode ray tube. The device was invented more or less simultaneously in Germany, Britain and several other countries which were concerned with air defence. The story of radar, or R.D.F. as it was originally called in England, has been frequently told. But one feature is worth recalling here. The successful development of centrimetric radar, using radio waves of only ten centimetres wavelength generated by a cavity magnetron was a uniquely British achievement. So too was the development of the technique known as 'operational research'.

These things apart, and their civilian applications should not be underrated, the British experience in technology over this period has been or is thought to have been, one of pioneering effort followed usually by the dismal experience of seeing the fruits of one's ideas exploited profitably by foreigners. Politicians have caught, and contributed to the mood. We have the ideas: the foreigners exploit them. It is an unjust and unworthy cry.

It is unjust for as we have pointed out even if foreigners do exploit British inventions the difficulties of the development stage should never be underestimated: it is one thing to have an airy-fairy idea, it is another thing to make it actually work. The story of Thomas Newcomen should have convinced us of this. It is also unjust in that it glosses over the probability that the foreigner may well have had the original idea at the same time. Indeed most modern inventions and scientific discoveries are made simultaneously and independently by two or more people. The reasons for this are simple and almost obvious. So many engineers and scientists are at work to-day, publication is so easy and discussions are so frequent that possibilities are always being canvassed as the progress of technology and science makes them feasible.

It is also unworthy in that it entirely ignores the importance of free communication in technology. We Europeans first learned our technology, in the forms of technics, from the Arabs to whom we are indebted not only for the inventions they themselves made but for the role they played as transmitters of ideas and inventions from China and India. This debt can only be repaid by ensuring that always in the future and in all circumstances technological inventions and ideas are freely communicated to all who can use or develop them. This does not exclude those whom we may choose to think are our rivals.

The complaint that other people come and steal our inventions is by no means peculiarly British. As we have already pointed out it has been made by Frenchmen and Russians; it was uttered by Germans in the days when their technology was still labouring to establish itself and no doubt it will be made by other peoples in the future. It reflects, as we have suggested, not so much a real state of affairs as a failure on the part of the complainant to understand the nature of technological progress and of his society or country to achieve a judicious balance between science and invention on the one hand and economic progress on the other.

The complaint has not, to my knowledge, ever been made by Japanese or American nationals. But the Japanese, who have now outgrown the imitative phase which was the unavoidable precursor of the establishment of a modern home-based technology, have hardly had time to appreciate the experience of a disharmony between inventiveness and scientific creativity on the one hand and a lethargic economy on the other. Of the Americans, who are perhaps the favourite butt of complaints of the sort deplored, it need only be said that no nation in the world has ever contributed so much to disinterested learning and the pursuit of knowledge or inventions for their own sake. This does not require proof in statistical terms; it is one of the evident facts of the twentieth century.

From the point of view of the historian of technology, as of the historian of science, the problem is in part an artificial one. It does not matter to him whether an engineer, inventor or applied scientist was an American, Briton, Frenchman, German, Japanese or Russian, or any other nationality, race, creed or colour. What matters to him is how the invention was conceived, developed and applied and what its consequences were. He must of course pay some attention to the social aspects, for the decline of a once creative community obviously must affect the development of technology and therefore the history of technology when it is written. That is to say, part of our understanding of the creative process in technology must include an appreciation of those factors which inhibit as well as those which encourage innovations. When a community ceases to be technologically creative, the historian must try to identify the causes of the decline even if, in the present state of knowledge, he cannot do much more than hazard guesses and put forward hypotheses.

The reasons why Germany ceased to be technologically active during the years 1630–1810* are many and complex;

* 1630 was the year in which Kepler died; 1810 the one in which the University of Berlin was founded. The approximate coincidence of the

and, significantly enough, what was true of German technology was equally true of German science. The two activities usually march hand in hand and what affects one affects the other. Under totally different economic and political conditions Great Britain experienced a not dissimilar failure in technological and scientific creativity between the heroic mid-nineteenth century period and the beginnings of the First World War. On the other hand Britain and France seem both to have known periods of scientific and technological activity accompanied by relatively slow economic progress; it is quite probable that Italy and Russia have had the same experiences. In brief, then, technological advance and economic progress are not always intimately linked although, of course, in the long run they are necessary to one another. Indeed a nation can get along merely by borrowing other people's ideas and without contributing anything much to the general fund of technological knowledge. Had this not been the case, then pre-medieval Europe, seventeenth-century England and modern Japan could never have acquired the wealth and the expertise that enabled them eventually to become technologically progressive. One must learn before one is able to teach others.

The proper indifference of the historian of technology to the nationality of particular inventors and technologists has tended to conceal the important fact that no nation has been very creative for more than an historically short period. Fortunately as each leader has flagged there has always been, up to now, a nation or nations to take over the torch. The diversity, inside a wider unity, of European culture—for Europe is the true home of technology—has made possible the continued growth of technology over the last seven hundred years.

SUMMARY AND CONCLUSIONS

Modern technology is often loudly proclaimed as perhaps the most powerful formative influence on our society, whether

---

Thirty Years' War with the beginning of German decline suggests an obvious causal relationship. But societies generally have remarkable powers of recuperation after even the most devastating of wars.

one approves of it or not. Exactly what meaning can be attached to the word technology is very doubtful for few writers and commentators bother to define their terms. One need not be a linguistic philosopher to deplore such carelessness and to be alarmed at the possible consequences of the generalisations that may proceed from it. The distinguished gentlemen, usually of the various literary professions, who denounce modern technology, put themselves in an untenable position. For without technology it is statistically certain that the majority of them would not be alive to denounce it. To be consistent, therefore, they should either 'opt out' by emigrating to some remote, desert island or take steps to reform those aspects of technology that they deplore; or at least identify those aspects so clearly and unambiguously that others can begin the process of reform, if such is found necessary.

To adapt a phrase of Wittgenstein's: what is technology is a matter of public opinion. This is not a rejection of the philosopher's claim to decide such questions but rather a recognition that both technology and science are changing so rapidly that old definitions are now out of date. In other words, in the present situation the historian and the philosopher require the aid of the informed sociologist. If public opinion holds that computers, space travel and nuclear reactors are great triumphs of science, then it is unwise to dismiss this as public ignorance of the difference between technology and science; it is more likely that the public feel that the two are indistinguishable and in this the public may well be right.

We may distinguish four stages in the development of technology. The first, preliminary stage coincided with the rise of European technics during the middle ages. With great moral and physical courage the miners, the navigators, the anatomists and other innovators of the period broke down a whole series of ancient and apparently universal tabus surrounding nature; and in doing so they immeasurably widened the horizons for European civilisation. Such, in fact, was the precocity, diversity and importance of medieval technics that one is inclined to believe that a mutation, philosophical or spiritual, had occurred, so that medieval civilisation was sharply and basically different from all those that had preceded it. The technics of the period were precocious in that they often ran far ahead of the very limited scientific knowledge available. They were diverse in the vast range of activities they were concerned with and they were important, not merely by virtue of their cumulative effects on the economic development of society but also because several medieval inventions decisively and fundamentally changed man's outlook on the world. Very few subsequent inventions had the same universal significance in this respect as did the weight-driven clock and the printing press. Modern society is still basically conditioned by these two medieval inventions!

The Chinese, as Dr Needham has demonstrated, are a very able people who have contributed an abundance of inventions to the common stock. Yet, in marked contrast to medieval Europe their culture became static and their inventive powers ossified. Could this possibly have been because, in a deferential society, administered by mandarins whose education was always severely literary, the simple criterion of utility was increasingly neglected? Many of the best known Chinese inventions seem to have been conspicuous rather more for ingenuity and even elegance than for practical utility. In contrast medieval Europe appears to have achieved a balance between utilitarian and what we might term 'idealistic' inventions. A good example of the latter was the weight-driven clock whose practical or cash value was only realised indirectly and in the long term. Indeed it might well be true that for technics (and technology) to prosper, both immediately useful and long-term, or even speculative inventions, must be encouraged.

The second important phase in the development of technology began in the early seventeenth century. Francis Bacon was the advocate and prophet, if not the originator, of the movement that was to establish technology—the word itself was first used, as we noted, in 1615—on the foundations of technics and science. With authority and clarity of style Bacon described two modes of technical innovation—what we have called 'empirical' invention and science-based invention. He examined the philosophical, psychological and social hindrances to technics and he went on to suggest ways in which the advance of technics might be expedited.

The complementary genius to Bacon's was that of Galileo.

For it was Galileo who, more than any other single individual, established modern physical science. His scientific procedures and his faith that the world order is ultimately mathematical can be traced back to the work of Archimedes and Plato respectively. But the mechanical philosophy that he professed owed a great deal to the achievements of medieval technics. According to the historian Collingwood:

> The Renaissance view of nature . . . (is partly) . . . based on human experience of designing and constructing machines. The Greeks and Romans were not machine users except to a very small extent; their catapults and water-clocks were not a prominent enough feature of their lives to affect the way in which they conceived the relation between themselves and the world. But by the sixteenth century . . . the printing press and the windmill, the pump and the pulley, the clock . . . were established features of daily life. Everyone understood the nature of a machine . . . It was an easy step to the proposition: as a clockmaker or a millwright is to a clock or mill, so is God to Nature.

Galilean method led, as we have shown, to the establishment of the science, or technology, of the strength of materials and to very important conclusions about the behaviour of machines. In particular Galileo made it possible for engineers to conceive of and to measure work and power in quantitative terms. This, in turn, enabled his disciples to establish the all-important idea of the quantifiable efficiency of machines or processes at the centre of technology, thus replacing the old normative criterion of how 'good' a machine was. Following Galileo it was possible to express the efficiency of a machine as a simple ratio between the energy of a moving agent on the one hand and the power the machine could develop on the other. It is possible that this revolution in technics, for it was nothing less than a revolution, owed something to the refinement of accountancy practices in the commercial houses of northern Italy. But this is a sector of history that has not yet been studied.

The third phase in the development of technology coincided with the movement known as the industrial revolution. It was characterised by the increasing use of Galilean procedures and the acceptance of the Baconian ideology. As science is a progressive, cumulative activity and as inventions in one industry stimulate inventions in others, so technology became more complex and several different types of technologist emerged in the course of the industrial revolution.

Besides the immemorial class of empirical inventors there appeared at least three other recognisable types. Characteristic of one group was Newcomen who was capable of using new scientific knowledge on which to base his important invention of the fire-engine. Another type was represented by Watt who used scientific, or experimental, method to investigate the unknown factors governing the performance of a fire-engine before making his invention of the separate condenser. (The work that Watt did in conceiving and developing the steam-engine would, in analogous circumstances today, be divided between a number of men: applied scientists, design, development and production engineers). The third and numerically the most important type was exemplified by the career of John Smeaton. The systematic procedure he used leads to evolutionary improvement but not to revolutionary new ideas. Smeaton's method enabled him to design and build Newcomen engines of maximum possible efficiency within the limits of the materials and ancillary gear available at the time; it could not, in principle, have led to the invention of a steam-engine with a separate condenser.* In the modern world Smeatonian method could account for the development of the piston aircraft engine to the peak of power and efficiency achieved just after the Second World War; it could not have led to the invention of the totally different jet engine.

If then at least four different modes of innovation corresponding to four characteristic types of technologist can be identified—and no doubt sub-divisions as well as new types might be recognised—there is another way in which we can classify inventors and technologists. There are those whose motives are simple and immediate personal gain: the Arkwrights and the Strutts, for example. Then there are men, like Watt and Trevithick, whose inventions are in response to actual needs but which take some time to develop and there-

---

* This is not to say that Smeaton could not have invented the separate condenser. It is an assertion that the rigorous application of his method logically precludes such inventions.

fore do not pay an immediate dividend. These are men in whose careers one may discern significant degrees of commitment to technology for its own sake. Finally we have the two small but important classes distinguished by men like Cayley and Sadi Carnot. The latter was a philosopher in his technology and his science; his was a broad, synoptic vision. Cayley had much more specific objectives, heavier-than-air flight being the most important. Nevertheless, to the extent that Carnot and Cayley were concerned with long term endeavours, they indicate that technology needs its prophets no less than its priests.

In short, it is misleading to talk of inventors and technologists as if they were all cast from one or two distinctive moulds so that they all appear as alike as guardsmen. Historical evidence suggests that if a civilisation is to remain technologically progressive, it must make it possible for a wide variety of talents to flourish. The Arkwrights and the Strutts are essential if the economy of a particular country is to prosper, but 'idealists' like Carnot and Cayley are also necessary in the long run and for the benefits of all men and all nations. The difficulty is that the present day organisation of technology, as of science, tends to iron out differences, to impose uniformity of ideas and to enforce conventional values. We all do, as Bernard Shaw once remarked, what society expects us to do. And it seems clear that society does not nowadays expect inventors, technologists or scientists to withdraw for years on end to develop their ideas and work out their problems in the way that was taken for granted by men like Carnot and Darwin. We cannot return to to the social forms of the eighteenth or nineteenth centuries; nor would we want to. But this does not solve the problem, nor absolve us from the obligation to face it squarely.

The fourth phase in the development of technology began about a hundred years ago with the establishment of industrial research laboratories, with the renaissance of German technology and science and with the rise of new technological powers: America and Japan. The conspicuous feature of the period was the rapid convergence of science and technology considered as social institutions; the virtual completion of the Baconian programme. Whitehead, whose life spanned this period, expressed the point very succinctly when he wrote: 'The greatest invention of the nineteenth century was the invention of the method of inventions'.

The phrase 'the method of inventions' refers, not to some esoteric methodological procedure but to the devising of institutions to ensure technological progress: the research laboratory, staffed by professional scientists, the design/ development department and technical sales and services. What had been, in the past, the highly distinctive features of occasional firms such as Boulton and Watt, or Sharp, Roberts, became increasingly commonplace in countries with advanced technologies. The main agents of this change have been the increasingly numerous classes of highly trained professional engineers and scientists that appeared first in Germany and America, later in France* and then, much later, in Britain.

In 1869 most men of science in England were amateurs. Joule and Darwin were at the head of a company whose various sources of income were inherited wealth, practices in law and medicine, church stipends and commerce and industry. Occasionally a university teacher would be found in this company and he would be virtually the only one who could claim professional status. The amateur scientists had, to compensate them for lack of public support, complete freedom in research, freedom from pressure to publish instant papers, no obligations to attend committees and academic boards or to dance attendance on government departments controlling research grants for post-graduate students.

This freedom coupled with the reasonable assurance that old Consols would never fall much below 99·5 per cent resulted in an atmosphere in which research could proceed in a leisurely way and with due philosophical detachment. The results had been not unimpressive. In Germany it was somewhat different. In that country most men of science were university professors and therefore technically civil servants. But in other respects they shared most, if not all, the freedoms of their British

* In the case of certain types of civil engineers it seems likely that France can claim priority over all other nations: we remember the 'polytechniciens' of the revolutionary period. But in the case of professionally trained mining engineers the honours must surely go to Germany for the famous mining academies established in the eighteenth century.

contemporaries. The international community of scientists was small and very select and it moved at a reasonable, human pace. As man is the measure of all things this was as it ought to have been. The Royal Society of London, with a total membership of about six or seven hundred included virtually all the recognisable scientists in Britain and carried very few of the passengers who had been such a prominent feature of the Society in the eighteenth century. Yet the numbers were too small and British amateurism was soon to prove no match for the professionalism of Continental scientists. It was only at the turn of the century that a class of professional scientists, comparable to the one that had existed in Germany for forty years or so, began to appear in Britain. But thereafter the triumph of the professional over the amateur was swift and complete.

In practically every way, apart from intellectual integrity, scientists of today are very different men from those who, up to the end of the nineteenth century, had created the already imposing structure of science. The modern scientist is a disciplined and highly specialised professional. It is, in short, arguable that we should use different words to distinguish the group of men to whom Galileo, Newton, Lavoisier, Clausius, Joule and Gibbs belonged from those who practice the contiguous sciences today. If we now have scientists, then those men might be better described as natural philosophers.

The essential difference lies in the administration, or rather organisation, of science. It is partly masked by the strong preconception that now exists that the real distinction is between 'pure', 'basic' or 'fundamental' science on the one hand and the applied sciences and technology on the other. It seems likely, however, that this distinction, which does not accord happily with the British empirical tradition, was imported into Britain only recently. The distinction between 'pure' science and technology was surely a mere reflection of the administrative distinction between the universities and the *technische hochschulen* in nineteenth-century Germany? Many Englishmen who, before 1914, returned home deeply—and very properly—impressed by German scientific achievements also brought back a tacit acceptance of the German distinction between 'pure' and applied science. Before about 1860 the expression 'pure' science was hardly to be found in British writings. Indeed to indicate the confusion it might have caused let us consider a short passage from Andrew Ure's writings:

> The best example of *pure* chemistry on self-acting principles which I have seen was in a manufacture of sulphuric acid, where the sulphur being kindled and properly set in train with the nitre, atmospheric air and water, carried on the process through a labyrinth of compartments and supplied the requisite heat of concentration till it brought forth a finished commercial product. The finest model of *mixed* chemistry is the five colour calico machine . . .

Ure was certainly sufficiently well-informed not to have used the expression 'pure' chemistry in the sense that he did had it been commonly used at that time to denote the kind of academic, non-applied chemistry to which it would refer today.

The distinction between pure and applied science would, one feels, have puzzled Kelvin and Rankine. Where, they might have asked, does it leave Sadi Carnot's *Réflexions sur la puissance motrice du feu*? Is it a work of 'pure' science or is it merely a manual for horny-handed engine builders that might, possibly, by accident as it were, have significant consequences for science? One has only to pose the question to realise how inappropriate it is, for science and technology have never been separate; they merge and in many places overlap. This has always been the case and it is even truer now than in the past.

The philosophical element exists among the present scientific population but it remains small and hard to identify. Possibly only the passage of time will enable us to recognise it. It is important to realise that, the 'pure science' doctrine notwithstanding, this element may well include scientists with a technological bent, applied scientists or technologists in their own right, latter-day Sadi Carnots. It is perfectly possible, likely perhaps, that an advance in computer technology, or in communications technology or even a drastic discovery in space travel, will have as profound an effect on our understanding of the world—our philosophy in short—as did the discoveries of the fifteenth-century navigators, the mechanical inventions of the middle ages and Carnot's conceptualisation of the ideal heat-engine.

All this, of course, has certain implications for the philosophy underlying our higher education. It is too easy to assume that young people who enter university science courses are hoping, in the main, to devote themselves to a life of 'pure' learning; or even that they conceive their disciplines as primarily concerned with 'pure science'. A tentative inquiry we have carried out suggests that a significant number regard a scientific training as a step towards a career in applied science and/or management. If this is generally true then our common assumptions about science, pure and applied, need revision; but on these points a good deal more information is required.

A point that is often made today is that young men are dissuaded from becoming engineers because of the ambiguity attaching to the name. Many parents and school teachers, it is suggested, look down on engineering as an occupation appropriate for wage-earning members of a trade union rather than for qualified professional people. But, if there is any truth in this, the inference drawn is surely the wrong one. Drivers of railway locomotives in America have long been called engineers: this does not seem to have discouraged young men from seeking admission to, for example, the Massachusetts Institute of Technology. On the other hand in most armies, engineers enjoy high prestige; the British army is no exception, for the Royal Engineers take precedence over all other troops save those of the Royal Household. If, indeed, confusion over names does discourage young people, or their parents or teachers, then the fault lies with the values of our society rather than with the simple word 'engineer'.

A more valid cause for concern is the fact that very few young women choose careers in technology. Astonishingly, for it seems to have escaped comment, many, if not most, young women, for all their vaunted independence and emancipation, still receive an education that would have been thought appropriate for the young ladies of a hundred and more years ago; that is, an education based on literary studies. Surely it is desirable from every point of view that more girls be encouraged and enabled to take up careers in technology and that many more, if not all, be given the opportunity to learn something about the technologies on which our civilisation now depends?

The rise of applied science, as a vital component of technology, can be appreciated if we compare the state of industry in 1900 with its state today. As we pointed out, seventy years ago many of the most important British industries had been established during the industrial revolution. Today, on the other hand, a large proportion of the working population is employed in industries that did not exist seventy years ago— indeed that were hardly dreamed of—for besides such new-comers as motors and aircraft, there are electronics and a wide range of electrical engineering products, fine chemicals and plastics, nuclear power and computers. All these new industries are based substantially on scientific research and depend to a considerable extent on continued applied science for their progress.

In the 1944 edition of his book, *Atomic Physics*, Professor Max Born had this to say about the properties of the element uranium:

> The process of fission is in fact accompanied by the emission of neutrons (either simultaneously, or later as disintegration products of the unstable fragments). Now these neutrons may be absorbed and produce fission in other neighbouring nuclei, neutrons being again emitted, so that an avalanche develops. If it were possible to produce such a chain reaction the enormously condensed nuclear energy could be utilised for practical purposes, as driving power for engines or as explosive for super-bombs.

In the following year, 1945, this chain reaction was publicy demonstrated. The subsequent development of nuclear energy for military and civil purposes indicates the effectiveness of science-based technology and the capacity of modern nations to muster and organise the very varied skills of technologists and scientists to achieve such remarkable results.

SOME RESIDUAL PROBLEMS

The success of mechanical technics and technology has been reflected in the imagery of mechanism in European thought over the last four hundred years or so. Clockwork in the seventeenth century and the heat-engine in the nineteenth have conditioned men's thoughts about nature; quite possibly the computer is on the point of doing the same thing for coming generations. But if we go back to the scientific revolution in the seventeenth century, we must wonder whether alternative non- or less mechanical forms of technology might not have been possible. Perhaps a very close reading of Francis Bacon and his contemporaries will indicate that a different form of technology, more in sympathy with the biological sciences, might have been in their minds? Men like William Gilbert, Johann Kepler, William Harvey and Bacon himself were not 'mechanical philosophers' and it is at least possible that technology might have taken a different course, more in accord with their ideals, had not Galileo, Descartes and Newton set the seal on the triumph of mechanism.

An alternative technology might well have avoided some of the stresses and distortions of modern society, but it could not, I think, have avoided the most intractable one, and it might have brought its own peculiar difficulties with it. There is no reason to believe that a different form of technology would have avoided the problems of specialisation and the division of labour; unless, that is, society was to remain at a subsistence level for the bulk of the population.

A pessimist might argue that practically all the 'big ideas' of the present day were propounded during the nineteenth century, or at least before 1914. Thus liberal, tory and social democracy, Marxism, psycho-analysis, women's suffrage, trade unionism, universal education, the popular motor car, the cinema, the popular press and a host of other features of contemporary society originated before 1914. In fact it might be said that since 1914 European civilisation has been relatively uncreative, judged by the standards of former times. Whether or not this generalisation is accepted—and, plainly, there is much to say on both sides—most people would point to technology as one instance of sustained and rapid progress. It is certainly true that since 1914 there have been revolutionary advances such as television, radar, the jet engine and nuclear power. But whether such advances come at more or less frequent intervals than they did in the nineteenth century is debatable. What is undeniable is that there has been an enormous expansion of evolutionary, or Smeatonian-type innovation. This expansion, which is a consequence of the extension of education allied to institutional specialisation, may have been effected at the expense of 'revolutionary' innovation. It is, after all, much easier to work as a member of a large team engaged on evolutionary development than it is to launch a revolutionary innovation of one's own.

Whatever the prospects for western technology as at present organised, it should not be forgotten that, on the world scale, it is still a very limited endeavour. Young technologists and applied scientists are trained to work in the research, development and production departments of advanced industrial and government organisations. This is, no doubt, as it should be and in accordance with the ambitions of the young men concerned. Nevertheless it is regrettable that the education of most young technologists does not include some understanding of the technics, with their attendant problems, of the greater part of mankind; technics that are still roughly at the level of the new stone age.

Attempts to raise the standards of living in 'undeveloped' countries have not, so far, been very successful. It is now realised that to institute a national airline, a university of technology, a steel works and a number of automated factories in an undeveloped society or nation does not automatically result in western style growth economy. Rather it further impoverishes the countryside and swells the city and town populations with unemployed craftsmen displaced by the injection of an alien technology. A reading of Babbage and Ure might have reminded those responsible that economic growth proceeds most rapidly in a society rich in technical skills and whose members have become accustomed to the general idea of innovation. 'Intermediate technology', which takes account of these things, now seems to be the best hope for the undeveloped world.

Some people will deny that the development of technology among the undeveloped nations is desirable, judged by the present state of Europe and North America. But it was not technology that instituted Auschwitz and Terezin, or for that matter destroyed Dresden. In all cases it was a series of military and political decisions based on certain philosophies that resulted in the crimes in question. In this matter we need look no further than original sin, for the religious explanation is not only adequate, it is obviously correct. What we can learn from such events is how an organised and highly specialised society makes it possible for large numbers of ordinary men, moderately respectable and no doubt charitable but also moderately cowardly where their own safety and the status of their families are concerned, to operate a system whose cumulative effect is a great evil. In such systems moral responsibility is diluted well below the tolerance levels of most men.

It is not technology that threatens our society and its values. For technology is itself a distinctive and dependent offspring of the philosophy, cosmology, and religion of that society. Technology is a highly refined instrument, although like all instruments it may be mis-applied. The problems are simply to ensure that it prospers—it was difficult enough to get it going in the first place!—and that it is used to attain what we are assured is a dignified and worthwhile life for civilised men.

At this point we come up against the problem of the impact of technology on the conditions of employment. What the mill and the mine were in the past, the assembly line is in the twentieth century. No observer can be happy about the man on the assembly line. Even after one has discarded the almost irresistible prejudice that one can arrange other people's lives to their better advantage and has made allowance for the undeniable fact that the assembly line worker enjoys luxuries and privileges that no Victorians enjoyed, or even dreamed of, there still remains the residual conviction that a man should be more than just one more component of an extremely complex, computer-controlled machine.

It seems probable that 'assembly line' situations arise, generally speaking, when technological advance outpaces slower-moving social philosophy. The advent of the Newcomen engine meant that many more women and children were employed in mines under conditions that were later judged to be deplorable. That particular problem was, however, overcome without jettisoning the technology of the heat-engine; so too, one hopes, will be the problem of the assembly line.

These and other social and economic problems of modern technology are common subjects of public discussion and are abundantly studied by social scientists. It is therefore appropriate to make a plea for the missing component, the study of the history of technology conceived not as a chronicle of gadgeteering, nor as a handmaid to economic history, but as an autonomous study closely related to the history of science, the history of ideas and to philosophy generally. Such a history of technology will not solve our problems; it will certainly not enable us to predict where, when, and how future inventions will be made. But it should give us a much-needed perspective that could eventually make us wiser in our judgments and more effective in our use of technology. Conversely, it will perhaps help us to understand the difficulties and avoid some of the dangers that must inevitably lie ahead.

Notes and Bibliography

Chapter One

NOTES

p. 3   Robert Lenoble, *op. cit.*

p. 27   A. C. Crombie, 'Physics and Astronomy: Introduction', in
*Atti del primo convegno internazionale di ricognizione della fonti per la
storia della scienza Italiana: Isecoli 14–16 Settembre, 1966* (Florence,
1967).

pp. 28–29   The claim has recently been made in R. A. Skelton and
others, *The Vinland Map and the Tartar Relation* (Yale, 1965) that
the Norsemen were, to judge by the evidence of the Vinland map
(c. 1440), probably the discoverers of America. Subsequent discussion has centred around the question whether the (correct)
representation of Greenland could possibly have been made
before the nineteenth century, much less before the time of
Columbus and also on the accuracy or otherwise of the positions
represented. 'America' is shown on the map as an island and this
is held to represent Labrador where there were almost certainly
pre-Columban Norse settlements. The discussion in fact seems to
be misplaced or at best premature: whether or not the Norsemen
settled in Labrador is irrelevant to the larger question of who
discovered America. Discovery, as commonly understood, implies
recognising the nature and general significance of what is discovered. In this respect there is no shred of evidence that the
Norsemen, or those who drew the Vinland map, had any appreciation of the significance of the 'island' of Labrador. Indeed, the
balance of probabilities would seem to be that if they gave it any
particular thought they regarded it as yet another Atlantic island
of the sort they were so familiar with: Great Britain and Ireland,
Iceland, Greenland, Spitzbergen, Jan Mayen and so on. The true
discovery of America surely requires the recognition that here was
a vast continent. This, on the evidence of the Vinland map, is
precisely what was neither done nor implied by the Norsemen.
Columbus may have thought after his first voyage that he had
reached the Indies by travelling West but his method of exploration was sound and although he was mistaken it was only a matter
of time before the true nature of his *discovery* was confirmed. If this

is denied and it is insisted that the first person or persons to reach America must be the true discoverers, then the credit belongs to the Eskimo or the Red Indians. But such an argument will I think be acceptable to very few!

BIBLIOGRAPHY

Ancient Technology and Science:

A. G. Drachmann, *The Mechanical Technology of Greek and Roman Antiquity* (Munksgaard, Copenhagen 1963).

R. J. Forbes, *Studies in Ancient Technology* (E. J. Brill, Leyden).

Vitruvius, *On Architecture* (Harvard University Press and Heinemann, 1962), 2 vols., Loeb Library.

Medieval Technology and Science:

A. C. Crombie, *Augustine to Galileo, the History of Science, A.D. 400–1650* (Mercury Books, London 1952; paper-back edition 1962).

Robert Lenoble, 'La Pensée Scientifique', in *Histoire de la science* (Editions Gallimard, Paris 1957), edited Maurice Daumas.

Lynn White, Jr., *Medieval Technology and Social Change* (Oxford University Press, 1962).

A. G. Keller, *A Theatre of Machines* (Chapman & Hall, London 1964).

Clockwork:

Carlo Cipolla, *Clocks and Culture, 1300–1700* (Collins, London 1967).

Ernest L. Edwardes, *Weight-driven Clocks of the Middle Ages and Renaissance* (John Sherratt, Altrincham 1965).

See also the various important papers by D. J. de Solla Price.

Printing:

Victor Scholderer, *Johann Gutenberg: the Inventor of Printing* (British Museum, London 1963). This is a short work with an excellent bibliography.

General:

Charles Singer, E. J. Holmyard, A. R. Hall and T. I. Williams, *The History of Technology* (Oxford University Press, 1958–1962), 5 vols.

T. K. Derry and T. I. Williams, *A Short History of Technology* (Oxford University Press, 1960).

George Sarton, *An Introduction to the History of Science* (Williams and Wilkins, for Carnegie Institution, Baltimore 1935–1962), 3 vols. in 4 parts.

A. C. Crombie (ed.), *Scientific Change* (Heinemann, London 1963).
Lewis Mumford, *Technics and Civilisation* (Routledge and Kegan Paul, London 1962).
Joseph Needham, *Science and Civilisation in China:* especially vol. IV: *Physics and Physical Technology* (Cambridge University Press, 1965).
Lynn Thorndike, *A History of Magic and Experimental Science* (Columbia University Press, New York 1958), 8 vols.
A. P. Usher, *A History of Mechanical Invention* (Harvard University Press, 1962).
Eugene S. Ferguson, *Bibliography of the History of Technology* (M.I.T., 1968)

Chapter Two

NOTES

p. 35   I am grateful to Dr P. M. Rattansi for telling me about this interesting comment by Wren.

pp. 37–41   The quotations are from Galileo's *On Machines.*

p. 42   Franz Reuleaux, *The Kinematics of Machinery* (London 1876, Dover edition 1963), p. 9.

p. 44   The quotation is from Galileo's *Two New Sciences.*

p. 46   A. N. Whitehead, *Science and the Modern World* (Cambridge University Press, 1925).

p. 50   W. P. D. Wightman, *The Growth of Scientific Ideas* (Oliver and Boyd, Edinburgh 1962) gives a good account of this famous experiment. See also Isaac Newton, *Opticks* (London 1704; Dover edition 1952).

pp. 54–55   Otto von Guericke, *Experimenta nova Magdeburgica de vacuo spatio* (Reprint, Otto Zeller, Aalen 1962).

p. 57   Daniel Defoe, *A Plan of the English Commerce* (Reprint, Oxford University Press, 1927). See also A. P. Wadsworth and Julia Mann, *The Cotton Trade and Industrial Lancashire, 1600–1780* (Manchester University Press, 1965). Wadsworth and Mann called attention to Defoe's comments.

BIBLIOGRAPHY

Francis Bacon:

F. Bacon, *The Advancement of Learning* (Everyman edition).

There is a large number of specialised studies dealing with Bacon's works. Few of them are important from the point of view of the historian of technology. Benjamin Farrington has written two short, sympathetic essays: *Francis Bacon, Philosopher of Industrial Science*

(Lawrence & Wishart, London 1949), and *Francis Bacon, Pioneer of Planned Science* (Weidenfeld & Nicholson, London 1963). Lynn Thorndike's essay (*op. cit.*, vol. VII, pp. 63–88) is anything but sympathetic.

Galileo:

Many of Galileo's most important works—*Two New Sciences; Two World Systems; Il saggiatore*, etc.—are available in paper-back editions. The best contemporary study is still the late Alexandre Koyré's *Etudes galiléennes* (Hermann et Cie, Paris 1939), 3 vols. Most regrettably there has been no extended study of Galileo's contributions to technology. The best source in this respect is, *Galileo: On Motion and On Mechanics* (Wisconsin University Press, 1960), edited by I. E. Drabkin and Stillman Drake. The essays and comments by the editors are particularly useful.

Arthur Koestler, *The Sleepwalkers* (Hutchinson, London 1959). A hostile but lucid critique of Galileo's ideas.

The Scientific Revolution of the Seventeenth Century:

This has been covered by a number of works. Among the best are A. R. Hall's *The Scientific Revolution* (Longman's, London 1962), A. C. Crombie, *op. cit.*, E. A. Burtt, *The Metaphysical Foundations of Modern Physical Science* (Routledge and Kegan Paul, London 1922), T. S. Kuhn, *The Copernican Revolution* (Harvard University Press, 1957) and *The Structure of Scientific Revolutions* (Chicago University Press, 1964), A. Wolf, *A History of Science, Technology and Philosophy in the XVIth and XVIIth Centuries* (George Allen & Unwin, London 1962).

Social and Intellectual Factors:

W. E. H. Lecky, *A History of . . . Rationalism in Europe* (Watts, London 1946). This gives a useful account of various seventeenth-century superstitions. More recent works include Christopher Hill's *Intellectual Origins of the English Revolution* (Oxford University Press, 1965) and the important papers by P. M. Rattansi in *Notes and Records of the Royal Society* and elsewhere.

The extensive bibliography of Newtonian studies that has grown up in recent years is of relatively little importance for the history of technology.

The New World View:

E. G. R. Taylor, *The Mathematical Practitioners of Tudor and Stuart England, 1485–1714* (Cambridge University Press, 1969), and *The*

Mathematical Practitioners of Hanoverian England, 1714–1840 (Cambridge University Press, 1966). These are invaluable reference works. Lt. Cdr. D. W. Waters, R.N. (Retd.), *The Art of Navigation in Elizabethan and Early Stuart Times* (Hollis & Carter, London 1958). This is a work of major scholarship in the history of technology.

Chapter Three

NOTES

p. 62 Parent's paper was published in *Histoire et Memoires de l'Academie Royale des Sciences* (1704). Accounts of it were subsequently given by Belidor, Desaguliers, Maclaurin, etc. The theory was accepted by certain English writers, like Olinthus Gregory and Thomas Tredgold, up to the second decade of the nineteenth century even though it had been thoroughly discredited theoretically, experimentally, practically and, one might add, *economically!*

p. 65 The quotation is from Dean Swift's *Gulliver's Travels*. 'A Voyage to Laputa'.

p. 69 Isaac Potter must have been that Englishman who, according to Gabriel Jars, arranged with the Schemnitz mines to be remunerated for his services by being given the net savings resulting from the use of Newcomen engines over the previous source of power (horses) for a limited time. This was a practice later used by James Watt and commonly believed to have been invented by him.

p. 70 J. T. Desaguliers, *op. cit.*, Vol. 2.

p. 79 J. T. Desaguliers, *op. cit.*, Vol. 2.

p. 80 John Smeaton's paper was published in *Philosophical Transactions of the Royal Society* (1759).

p. 83 The quotation from John Farey is from his *Treatise on the Steam Engine*.

pp. 88–89 See John Robison's *A System of Mechanical Philosophy* (Edinburgh 1822), Vol. 2. Also J. P. Muirhead's *Life of James Watt* (London 1859).

p. 93 See the paper by Eugene S. Ferguson, 'Kinematics of Mechanism from the Time of Watt', *Selected Smithsonian Papers* (1962), No. 27.

p. 93 Letter from James Watt to Dr William Small, 28th May 1769. In Boulton and Watt Collection, Central Reference Library, Birmingham.

pp. 94–107 For the ideas put forward in these pages I am much indebted to my friends and colleagues, Drs R. L. Hills and A. J. Pacey and to the works of two former postgraduate students in my Department, Mr M. C. Egerton and Mr S. B. Smith.

Copies of the 1769 and 1777 agreements between Arkwright and his partners were very kindly given to my Department and to UMIST by Mr E. G. Holiday some years ago. Mr Holiday is a direct descendant of John Smalley. See Dr R. S. Fitton's article in *Advance* (April, 1967) 11, p. 10.

The number of Arkwright's patent of 1769 was 931 and that of James Watt in the same year, curiously enough, was 913.

BIBLIOGRAPHY

The Steam-Engine:

John Farey, *A Treatise on the Steam Engine* (London 1827). This is, in many ways, still the most comprehensive and scholarly work on the history of the steam-engine. A second volume, dealing mainly with the later development of the high-pressure engine, reached the stage of page-proof but was never actually published. The proofs are at the Patents Office Library, London. This work merits careful editing, indexing and publishing.

L. T. C. Rolt, *Thomas Newcomen, the Pre-history of the Steam-engine* (David & Charles, Dawlish and London 1963). This book gives a readable account of what is known about Thomas Newcomen and the early history of the atmospheric engine.

E. Robinson and D. McKie, *Partners in Science: Letters of J. Watt and J. Black* (Constable, London 1970).

E. Robinson and A. E. Musson, *James Watt and the Steam Revolution* (Adams & Dart, London 1969).

D. S. L. Cardwell, *Steam Power in the Eighteenth Century* (Sheed & Ward, London 1963), and *Watt to Clausius; the Rise of Thermodynamics in the Early Industrial Age* (Heinemann, London 1971). Both books give accounts of the scientific background to the development of the Steam-engine.

Mr Trevor Turner is writing the biography of John Smeaton.

Professor Eric Robinson is writing the biography of James Watt.

Textiles:

R. S. Fitton and A. P. Wadsworth, *The Strutts and the Arkwrights, 1758–1830* (Manchester University Press, 1964).

C. Aspin and S. D. Chapman, *James Hargreaves and the Spinning Jenny* (Helmshore Local History Society, 1964).

H. Catling, *The Spinning Mule* (David & Charles, Newton Abbot 1969).

For an account of the complex problems of applying power to textile machinery and the ways in which these problems were solved, see R. L. Hills, *Power in the Industrial Revolution* (Manchester University Press, 1969).

There is, surprisingly, no published volume dealing with the development of water-power technology during the eighteenth century: the very sinews of the industrial revolution!

General:

Robert E. Schofield, *The Lunar Society of Birmingham* (Oxford University Press, 1963), gives an account of the scientific and technological circle to which James Watt and Matthew Boulton belonged.

Professor Arnold Thackray is undertaking a study of the early history of the Manchester Literary and Philosophical Society and of the scientists and technologists associated with it.

Chapter Four

NOTES

p. 103   R. L. Hills, *op. cit.*

p. 112   William Jackson, *The Four Ages* (London 1798).

p. 112   I am indebted to Mr Trevor Turner for this interesting quotation.

p. 113   J. Laurence Pritchard, *op. cit.*, p. 36.

p. 113   In *Annales de Chimie*, xxxi (Year 7—1799), p. 26, a letter from Joseph Montgolfier to General Meusnier is quoted in which it is suggested that soldiers, dropped by parachutes from balloons, might be made to 'rain down' on hostile cities. This appears to be the first reference to parachute troops!

p. 125   G. S. Laird Clowes, *op. cit.*

p. 125   For French contributions to civil engineering at this time, see Hans Straub, *op. cit.*, p. 173.

p. 127   Messrs. A. E. Musson and E. Robinson have recently claimed that science was extensively used in the industrial revolution (*Science and the Industrial Revolution* (Manchester University Press, 1970)). They base their claim on industrial records and on the science classes held in the industrial areas. In order to avoid confusion, therefore, we must clarify what is meant by science in this context.

It cannot mean continued, systematic research, carried out by professional scientists. Such science could not begin until about 1870 when a supply of professional scientists became available, first of all in Germany, later in various other countries (see D. S. L. Cardwell, *The Organisation of Science in England* (Heinemann, London, 1957)).

Similarly it cannot refer to the leading ideas and the key researches of the period. For the mechanics of Lagrange, the development of Voltaic electricity, the physics of gases, sidereal astronomy, the analytical theory of heat and the undulatory theory of light played little or no part in the industrial revolution. And, as we have seen, France was the leading scientific nation during the period of England's industrial revolution.

But if science is taken to mean established knowledge, the doctrines of Galileo, Mariotte, Hooke and subsequent mechanical philosophers, then the thesis becomes plausible and important. We infer that as more new machines and processes were invented, so the usefulness of established scientific knowledge would be increasingly apparent.

At the same time the industrial revolution undoubtedly helped to set the scene for the new and very characteristic sciences of the nineteenth century. But this is another story.

p. 128–9 The figures are taken from the Lean brothers' *Historical Account*.

p. 129 *et seq.* The quotations are from Carnot's *Réflexions sur la puissance motrice du feu.*

p. 133 The representation of the Carnot cycle by means of an indicator diagram was due to Emile Clapeyron. Figure 28 should be compared with Figure 30.

p. 138 There is an account of Burdin's turbine in *Annales de chimie et de physique*, xxvi (1824), p. 207.

BIBLIOGRAPHY

The Steam-Engine:

Francis Trevithick, *Life of Richard Trevithick* (London 1872). See also the modern biographies by H. W. Dickinson and A. F. Titley and by L. T. C. Rolt.

For Cornish engines generally, see D. Bradford Barton, *The Cornish Engine*, T. R. Harris, *Arthur Woolf, the Cornish Engineer*, and Edmund Vale, *The Harveys of Hayle* (all by D. Bradford Barton, Truro 1966). Lean's reports were summarised and published in one volume by his sons: *Historical Statement of the Steam-engines in Cornwall* (London 1839).

Later History of the Firm of Boulton & Watt:

Economic and organisational aspects were discussed in (Sir) Eric Roll's *An Early Experiment in Industrial Organisation* (Frank Cass, London 1968 edition), and John Lord's *Capital and Steam Power* (Frank Cass, London 1966 edition).

General Engineering:

A. F. Burstall, *A History of Mechanical Engineering* (Faber & Faber, London 1963).

P. J. Booker, *A History of Engineering Drawing* (Chatto & Windus, London 1963).

G. S. Laird Clowes, *Sailing Ships, their History and Development* (H.M.S.O., Science Museum Publications).

Friedrich Klemm, *A History of Western Technology* (Allen & Unwin, London 1959).

Ian McNeil, *Joseph Bramah* (David and Charles, Newton Abbot 1968).

L. T. C. Rolt, *Tools for the Job* (Batsford, London 1965).

Hans Straub, *History of Civil Engineering* (Leonard Hill, London 1960).

R. S. Woodbury, *History of the Lathe to 1850* (1961), *History of the Gear-cutting Machine* (1958), *History of the Milling Machine* (1959) (all by M.I.T. Press).

Science and Technology in France:

M. P. Crosland, *The Society of Arcueil* (Heinemann, London 1967), and Robert Fox, *The Caloric Theory of Gases from Lavoisier to Regnault* (Clarendon Press, Oxford 1971).

Thermodynamics:

S. Carnot, *Réflexions sur la puissance motrice du feu* (facsimile edition Paris 1953). This work was translated some seventy years ago by R. H. Thurston. The translation was republished a few years ago by Dover Books together with certain other papers on thermodynamics and a most useful introduction by Professor Eric Mendoza.

General:

Abraham Rees' *Cyclopaedia* (1819) and the third edition of the *Encyclopaedia Britannica* (1797), are invaluable sources for this period.

Aviation:

J. Laurence Pritchard, *Sir George Cayley: Inventor of the Aeroplane* (Max Parrish, London 1961).

Chapter Five

NOTES

p. 140   P. and E. Morrison, *op. cit.*

p. 155   R. J. E. Clausius, First Memoir, 'On the Moving Force of Heat' (Berlin 1850).

pp. 158–60   R. J. E. Clausius, Ninth Memoir, 'Convenient Forms of the Fundamental Equations' (Zurich 1865).

p. 162   W. S. Jevons, *op. cit.*, expounds this philosophy forcefully and clearly.

p. 163   Charles Babbage, *The Great Exposition of 1851.*

pp. 165–6   Alphonse Beau de Rochas, *Nouvelles récherches* (Paris 1862).

pp. 168–9   Rudolf Diesel, *Theory and Construction of a Rational Heat Motor* (London 1894).

p. 168   The classical experiment carried out by J. P. Joule proved that, in an 'isothermal' expansion all the heat absorbed is converted into external work.

p. 178   The quotations are from *The Scientific Papers of J. Clerk Maxwell*, edited by W. D. Niven (Cambridge University Press, 1890), pp. 477 and 486.

Maxwell left behind a whole series of unsolved but fruitful problems. For example, he did not explain what his 'elementary particles' of electricity were supposed to be and he did not account for the origin of light, considered as an electromagnetic radiation. These, and other questions, were answered most profitably by succeeding generations; among the most interesting consequence being the theories of relativity. For accounts of these developments see Whittaker, *op. cit.*, and *Historical Studies in the Physical Sciences* (University of Pennsylvania Press, 1969), vol. 1, edited by Russel McCormmach.

In our discussion of Maxwell's theory we cannot be concerned with the modifications imposed on it by the theory of relativity. Our concern is with the immediate practical consequences.

pp. 182–3   The quotation is from Henri Poincaré and Frederick K. Vreeland, *Maxwell's Theory and Wireless Telegraphy* (London and New York, 1905), p. 71.

p. 185   Heinrich Hertz, *Electric Waves* (London and New York, 1893), translated by D. E. Jones.

p. 187   Charles Susskind, *op. cit.*
p. 194   H. J. Habbakuk, *American and British Technology in the Nineteenth Century* (Cambridge University Press, 1962).

BIBLIOGRAPHY

General:

Charles Babbage, *Reflections on the Decline of Science in England* (London 1830); *On the Economy of Machines and Manufactures* (London 1833, reprint Kelley, New York 1963); *Charles Babbage . . . Selected Writings*, edited by P. and E. Morrison (Dover Books, New York 1961); *The Great Exposition of 1851* (London 1851, reprint Frank Cass, London 1968).

Andrew Ure, *The Philosophy of Manufactures* (London 1835, reprint Frank Cass, London 1965).

W. O. Henderson, *The Industrial Revolution on the Continent* (Frank Cass, London 1967).

Sir John Clapham, *The Economic Development of France and Germany* (Cambridge University Press, 1936 edition).

J. B. Bury, *The Idea of Progress* (London 1922, reprint Dover Books, New York 1955).

W. S. Jevons, *The Coal Question* (London 1862 edition).

Educational and Institutional:

Dame Mabel Tylecote, *The Mechanics' Institutes of Lancashire and Cheshire* (Manchester University Press, 1957)

D. S. L. Cardwell, *The Organisation of Science in England* (Heinemann, London 1957).

Thermodynamics and the Heat-Engine:

Max Planck, *A Treatise on Thermodynamics* (London 1926, reprint Dover Books, New York n.d.).

D. S. L. Cardwell, *Watt to Clausius* (*op. cit.*).

There is, regrettably, no book in English that gives a fair and adequate account of the work of R. J. E. Clausius.

Arthur Evans, *A History of the Oil Engine* (Sampson Low, Marston, London n.d.).

Dugald Clerk, *The Gas and Oil Engine* (London 1896). This early standard work contains some useful historical material.

Machine Tools:

L. T. C. Rolt, *Tools for the Job* (*op. cit.*).

The Official Catalogue of the Great Exhibition of 1851; The Exhibited Machinery of 1862; Report of the Paris Universal Exhibition, vol. IV (H.M.S.O., London 1868).

Singer, Holmyard, Hall and Williams, *op. cit.*

Derry and Williams, *op. cit.*

Textiles:

Evan Leigh, *The Science of Modern Cotton Spinning* (Manchester 1882), 2 vols.

Electricity:

Percy Dunsheath, *A History of Electrical Engineering* (Faber & Faber, London 1962).

Sir Edmund Whittaker, *A History of the Theories of Aether and Electricity* (Thomas Nelson, London 1961–2), 2 vols. This is a work of major scholarship; the account given of Hertz's work is, however, brief and there is no mention of Marconi although there is a reference to D. E. Hughes.

Charles Susskind, *Popov and the Beginnings of Radiotelegraphy* (San Francisco Press, 1962).

There is no detailed work that describes the emergence of the technology of radio-communications from the science of Maxwell and Hertz.

Foreign Investment in British Industry in the 19th and 20th Centuries:

John H. Dunning, *American Investment in British Manufacturing Industry* (George Allen & Unwin, London 1958).

Chapter Six

NOTES

p. 197   The classical atom was supposed to be a universal building block of nature, common to all substances. A Daltonian atom, on the other hand, was present only in the particular element of which it was a building block, or in compounds containing that element. In this sense Dalton's atom represented a break with the traditional atomism inherited from the Greeks and subscribed to, nominally at least, by the majority of scientists from the seventeenth century onwards. Electrons, protons, neutrons, etc. are, on the other hand, true universal building blocks, common to all elements and compounds. Strictly speaking they should therefore be called 'atoms' while the Daltonian particles should be called corpuscles or concretions.

p. 199–200   John B. Rae, *The American Automobile*, p. 6.

p. 205   I am relying for the information about Archdale's on my recollections of a conversation I had a number of years ago with the late Mr Fred Archdale, at that time Managing Director of James Archdale, Ltd.

p. 205   Charles Singer, 'Technology and History', L. T. Hobhouse Memorial Lecture (London 1952), p. 8.

p. 207   Sir Frank Whittle, *op. cit.*, facing p. 97.

p. 207   The initials 'R.D.F.' could easily be confused with those of the much older and well-known technique of radio direction-finding (D/F). Such confusion was obviously useful (cf. the first World War 'tank').

p. 207   The basic technique of radar was first developed as a scientific method of finding the height of radio reflecting layers, such as the Kennelly-Heaviside layer (p. 188 above) and the Appleton layer, in the upper atmosphere (1926 onwards). Radar was thus a development of established scientific techniques.

pp. 209–10   Chemistry has almost always been a distinctively German science, but during the eighteenth century the main contributions came from France, Holland, Scotland, Sweden and England.

In England in the second half of the nineteenth century the pursuit of chemistry had almost ceased, the only (relatively) prominent figures being Roscoe, Frankland and Crookes. They should be compared with Davy, Dalton, Wollaston, the young Faraday and Prout at the beginning of the century. The study of thermodynamics had virtually ceased, as we have shown, and had it not been for the efforts of the small Cambridge mathematics group, physics too might have stopped.

p. 213   R. G. Collingwood, *The Idea of Nature* (Oxford University Press, 1945), pp. 8–9.

p. 216   A. N. Whitehead, *Science and the Modern World*, Chapter 6.

p. 218   Andrew Ure, *The Philosophy of Manufactures*, p. 2.

The distinction between 'pure' science and the other sort would also have puzzled such writers as Desaguliers and Robison, who included accounts of the steam-engine in their text-books of 'natural philosophy'. Very few British natural philosophers held themselves aloof from technology.

p. 220   Max Born, *Atomic Physics* (Blackie, London 1944), p. 195.

BIBLIOGRAPHY

There is an abundance of books dealing with the recent history of technology. These include some excellent company histories, personal accounts by distinguished technologists and a variety of specialised monographs. Only a small selection of these can be mentioned.

John B. Rae, *The American Automobile* (University of Chicago Press, 1965), and *Climb to Greatness: the American Aircraft Industry, 1920–1960* (M.I.T. Press, 1968).

W. H. G. Armytage, *The Rise of the Technocrats* (Routledge and Kegan Paul, London 1965), and *The Social History of Engineering* (Faber & Faber, London 1966).

Sir Frank Whittle, *Jet, the Story of a Pioneer* (Muller, London 1953).

Sir Robert Watson-Watt, *Three Steps to Victory* (Odhams Press, London 1958).

John Jewkes, David Sawers and Richard Stillerman, *The Sources of Invention* (Macmillan, London 1969 edition). The authors argue very strongly that major inventions come from individual inventors and not, usually, from large-scale organisations.

C. F. Carter and B. R. Williams, *Industry and Technical Progress* (Oxford University Press, 1957). The report of an intensive attempt to identify the factors that impeded the application of new scientific ideas in British industry.

Frank G. Wollard, *Principles of Mass and Flow Production* (Iliffe, London 1954). A short and very readable book by one of the men responsible for the first transfer machines in 1924.

John Diebold, *Automation: the Advent of the Automatic Factory* (Van Nostrand, New York 1952).

F. J. M. Laver, *Introducing Computers* (H.M.S.O., London 1965).

Norbert Wiener, *The Human Use of Human Beings* (Eyre & Spottiswoode, London 1950). The reflections of a distinguished mathematician on the possible consequences of the control systems to whose development he contributed.

Derek Hudson and Kenneth Luckhurst, *The Royal Society of Arts, 1754–1954* (John Murray, London 1954).

J. E. Gerstl and S. P. Hutton, *Engineers: the Anatomy of a Profession* (Tavistock Publications, London 1966).

D. G. Christopherson, *The Engineer in the University* (English Universities Press, London 1967).

Leslie Holliday (ed.), *The Integration of Technologies* (Hutchinson, London 1966).

William H. Davenport and Daniel Rosenthal (eds.), *Engineering: its Role and Function in Human Society* (Pergamon Press, London and New York 1967).

Index